Time-Domain Studies of the Andromeda Galaxy

AAS | IOP Astronomy

About the program:

AAS-IOP Astronomy ebooks is the official book program of the American Astronomical Society (AAS), and aims to share in depth the most fascinating areas of astronomy, astrophysics, solar physics and planetary science. The program includes publications in the following topics:

- GALAXIES AND COSMOLOGY
- INTERSTELLAR MATTER AND THE LOCAL UNIVERSE
- STARS AND STELLAR PHYSICS
- EDUCATION, OUTREACH AND HERITAGE
- HIGH-ENERGY PHENOMENA AND FUNDAMENTAL PHYSICS
- THE SUN AND THE HELIOSPHERE
- THE SOLAR SYSTEM, EXOPLANETS, AND ASTROBIOLOGY
- INSTRUMENTATION, SOFTWARE, LABORATORY ASTROPHYSICS AND DATA

Books in the program range in level from short introductory texts on fast-moving areas, graduate and upper-level undergraduate textbooks, research monographs and practical handbooks.

For a complete list of published and forthcoming titles, please visit iopscience.org/books/aas.

About the American Astronomical Society

The American Astronomical Society (aas.org), established 1899, is the major organization of professional astronomers in North America. The membership (~7,000) also includes physicists, mathematicians, geologists, engineers and others whose research interests lie within the broad spectrum of subjects now comprising the contemporary astronomical sciences. The mission of the Society is to enhance and share humanity's scientific understanding of the universe.

Time-Domain Studies of the Andromeda Galaxy

Chien-Hsiu Lee
Subaru Telescope, National Astronomical Observatory of Japan, 650 N Aohoku Pl, Hilo, HI 96720, USA

IOP Publishing, Bristol, UK

ISBN 978-0-7503-1353-7 (ebook)
ISBN 978-0-7503-1354-4 (print)
ISBN 978-0-7503-1355-1 (mobi)

DOI 10.1088/2514-3433/ab3639

Version: 20190901

AAS–IOP Astronomy
ISSN 2514-3433 (online)
ISSN 2515-141X (print)

British Library Cataloguing-in-Publication Data: A catalogue record for this book is available from the British Library.

Published by IOP Publishing, wholly owned by The Institute of Physics, London

IOP Publishing, Temple Circus, Temple Way, Bristol, BS1 6HG, UK

US Office: IOP Publishing, Inc., 190 North Independence Mall West, Suite 601, Philadelphia, PA 19106, USA

To my family, without whom I could not chase the stars and the galaxies so far.

Contents

Preface

The advent of charge-coupled devices and the subsequent implementations of ultra-wide cameras provided a firm basis for large imaging surveys, which opened up a new window for time-domain astronomy. In the beginning, time-domain studies were focused on monitoring campaigns of selected dense stellar fields. This included the Milky Way (the Galactic bulge in particular) and its satellites the Magellanic Clouds. The monitoring campaigns also extended to other nearby galaxies such as the Andromeda and Triangulum galaxies, to name a few. However, with the advance of technologies, time-domain studies have been transformed by un-biased all-sky surveys, such as the All-Sky Automated Survey, Pan-STARRS, Palomar Transient Factory, Catalina Sky Surveys, to name a few. With the Large Synoptic Survey Telescope (LSST) on the horizon, there will be an avalanche of time-series data of the entire southern sky. In light of LSST, it is important to prepare future generation of astronomers with the knowledge and tools to digest the big data in a timely manner. In this book, I will use our closest neighboring spiral galaxy—the Andromeda galaxy—as an example, to walk the readers through the ideas and techniques of time-domain analysis. This includes using image subtraction to obtain high quality time-series photometry, determining periods of variables using periodogram and classifying transients and variables based on the shape of the light curves. I also highlight some science topics to showcase how we can use the time-domain data to shed light on the nature of the transient events and their progenitors. In addition, we can also utilize variable stars to facilitate our understanding of stellar physics and use them as distance, age, and metallicity indicators. Along with this book, I am also preparing an online database of the Andromeda galaxy time-series data gathered from archival observations, in combination with multi-wavelength observations ranging from X-ray to far-infrared, such that the readers can have hands-on experience of real data.

Chien-Hsiu Lee at Tucson, June 2019

Author biography

Chien-Hsiu Lee

Chien-Hsiu Lee is an assistant scientist at the National Optical Astronomy Observatory, USA. He earned his bachelor and master's degree in Taiwan and then pursued his Ph.D. studies at the International Max Planck Research School on Astrophysics in Germany. He has been involved in several large imaging surveys, including the Pan-STARRS, Palomar Transient Factory, Hyper Suprime-Cam Subaru Strategic Program, and Large Synoptic Survey Telescope. As an assistant scientist at NOAO, he spends the majority of his time for services in the Community Science and Data Center; he spends the rest of his time in research, especially on exoplanets, time-domain studies, and gravitational lensing.

Chapter 1

Introduction

"Like a candle seen through a horn"

—Simon Marius, 1612

1.1 Basis Properties and Early Observations of the Andromeda Galaxy

The Andromeda galaxy (hereafter M31) is an intermediate spiral galaxy[1] spanning approximately 180 × 60 arcminutes (de Vaucouleurs et al. 1991) on the sky (see Figure 1.1). It is visible with the naked eye, thus was already known in the pre-telescope era. It was recorded in star catalogs as early as the 10th century by the Persian astronomer Abd al-Rahman al-Sufi and was also included in Charles Messier's catalog as M31. German astronomer Simon Marius was the first one to observe the M31 through eyepieces of a telescope in 1612. He described M31 as "a candle seen through a horn." Due to its nebulous appearance, M31 had long been thought to be a nebula in the Milky Way. In 1885, the supernova S Andromeda (which was mistakenly considered as a nova at that time) outshone the entire M31 nebula (Hartwig 1885), stimulating interest in this great nebula and subsequent follow-up observations to search for novae in M31 (Hubble 1929). In the meantime, several lines of evidence pointed to the galaxy nature of M31. For example, Julius Scheiner (1899) noticed that the spectrum of M31 shows solar-like absorptions, and concluded that M31 must be a system of stars. Vesto Slipher (1913) measured the radial velocity of M31 to be -300 km s^{-1}. This is difficult to reconcile with the claim that M31 is a nebula within the Milky Way. These pieces of evidence led to the Great Debate in 1920 between Harlow Shapley and Heber Curtis. Harlow Shapley argued that if M31 were not a spiral nebula, then S Andromeda would have outshone the entire galaxy, which was considered an impossible energy output at that time. On the

[1] M31 is classified as "Sb" type according to Hubble's taxonomy (Hubble 1926) or "SA(s)b" according to the third reference catalog of bright galaxies (de Vaucouleurs et al. 1991).

doi:10.1088/2514-3433/ab3639ch1

Figure 1.1. Optical three-color image of the Andromeda galaxy, as well as its companion satellite galaxies M32 and NGC 205. Image Credit: Dr. Wei-Hao Wang, ASIAA.

other hand, Heber Curtis regarded M31 as an "island universe." He pointed out that there were more novae in M31 than in the Milky Way. If M31 is not a separate galaxy, why would more novae occur in such a small region. The extra-galactic nature of M31 was established by Edwin Hubble in the 1920s. At that time, Hubble first conducted detail studies of M31 variables and determined its distance using Cepheids, settling the Great Debate between Harlow Shapely and Heber Curtis.

At a distance of 752 ± 27 kpc, or a distance modulus of 24.38 ± 0.06 (statistical) ± 0.03 (systematic) mag (Riess et al. 2012), M31 is the closest spiral galaxy to the Milky Way. Its proximity enables us to resolve its stellar content and provides an ideal testbed for stellar physics, galaxy formation and evolution, as well as cosmology. Modern, systematic observations of M31 can be dated back to the pioneering work by Edwin Hubble in the 1920s. Using 350 photo plates taken with the 60 and 100 inch telescopes at the Mount Wilson Observatory, Edwin Hubble (1929) presented detail studies of M31 variables (including Cepheids and novae) and established its extra-galactic nature. During World War II, Baade (1944) at the Mount Wilson Observatory took the advantage of wartime blackout and resolved the central region of M31 and its companion galaxies into stars for the first time. This led him to the conclusion of two distinct stellar populations, i.e., Population I (Pop I) from young stars and Population II (Pop II) from old stars, in these galaxies. Furthermore, based on the idea of two distinct stellar populations, he also discovered the two types of Cepheids, i.e., classical Cepheids from Pop I star and Type II Cepheids from Pop II stars (Baade 1956). This resolved the discrepancy of Cepheid Period-Luminosity relation between the Milky Way (classical Cephieds) and globular clusters (Type II Cepheids). In the 1970s, Rubin used the advanced spectrometer built by fellow astronomer Ford to study the rotation curve of M31 (Rubin & Ford 1970). To their surprise, the rotation rate at the outskirts of M31 does not drop as expected, suggesting a large reservoir of unseen mass. This provides another piece of evidence of the "dark matter," proposed and coined by Fritz Zwicky (1933).

1.2 Modern M31 Observations with Digitized Detector

Since then, our understanding of M31 has been dramatically changed by the advent of charge-coupled devices (CCDs) and wide-field cameras. There are several waves of M31 surveys with different aims. The first wave is to search for distance indicators (e.g., Cepheids and eclipsing binaries) to pin down the distance of M31. For example, the DIRECT project[2] (Macri 2004) made use of the 1.2 m telescope at the F. L. Whipple Observatory (FLWO) and 1.3 m Michigan-Dartmouth-MIT (MDM) telescope, equipped with an 11 × 11 arcminutes imager, to observe M31 (as well as M33) for ~200 nights during 1996 September and 1999 November. This yields the identification of ~600 Cepheids, ~130 eclipsing binaries, and ~500 miscellaneous variables. In the meantime, there was also another M31 monitoring project to search for eclipsing binaries using the 0.5 × 0.5 degrees wide-field camera (WFCAM)

[2] http://www.astronomy.ohio-state.edu/~kstanek/CfA/DIRECT/.

mounted on the 2.5 m Isaac Newton Telescope, with the goal to determine the M31 distance better than 4% (Ribas et al. 2004). This WFCAM/INT project has discovered ~4000 variables in the 0.5 × 0.5 degrees field in the northern disk of M31, where ~440 are eclipsing binaries and ~420 are Cepheids. With extensive spectroscopic follow-up of two bright (V = 19–20 mag) eclipsing binaries, Vilardell et al. (2010) were able to determine the M31 distance at 4% level, the most precise independent distance estimate. This rendered M31 a distance anchor for cosmological distance ladder and helped pin down a 2.4% Hubble constant with local measurement (Riess et al. 2016).

1.3 Microlensing Surveys and Spin-Offs

The second wave of the M31 monitoring campaign is searching for microlensing events to constrain the fraction of massive compact objects such as dark matter in the halo of M31. While Albert Einstein (1936) first provided the formula of the gravitational lensing behavior of stellar mass objects serving as lenses, he concluded that such events are extremely rare and are unlikely to be observed. Bohdan Paczynski (1986) was the first one to investigate the probability (or optical depth) of such events; based on his calculations, while the event rate is relatively low (one in a million), it is still possible to identify such events with the advent of modern CCD detectors, provided that we can observe a very dense stellar region continuously. Inspired by Paczynski's seminal paper, several microlensing campaigns, including MACHO, EROS, OGLE, and MOA, started to observe the Galactic bulge and Magellanic Clouds to search for signs of massive compact halo objects. While the MACHO team reported the detection of 10 microlensing events and estimated 16% of the Milky Way halos is made up of massive compact objects with masses between 0.1 and 1 solar masses, the EROS and OGLE team suggested that self-lensing (stars in the Magellanic Clouds lensed by another star in the Magellanic Clouds) alone can fully account for all the microlensing events they found. These contradicting results can be partly attributed to our limited sight-line toward the Galactic bulge and the Magellanic Clouds, as well as the small number of microlensing events.

In this regard, the attention of microlensing searches have turned to M31, the closest spiral galaxy. This is because we can have multiple sight-lines across M31, including its bulge and disk that will help better constrain the self-lensing rates. Using the Vatican Advanced Technology Telescope (VATT), Crotts & Tomaney (1996) reported the first detection of microlensing events in M31. Since then, several microlensing monitoring campaigns of M31 have been initialized. Here, we focus on three of them. The first one is the POINT-AGAPE project[3] (Aurière et al. 2001), which made use of the WFCAM/INT to monitor two fields in the northern and southern disks of M31 between 1999 and 2001, with a total of ~180 epochs. The second one is the POMMES project, which utilized the 1 × 1 squared degree MegaCAM on-board the 3.6-m Canada–France–Hawaii Telescope to monitor two fields in the northern and southern disks of M31 between 2004 and 2005, with a total

[3] http://www.ing.iac.es/PR/SH/SH2006/agape.html.

of 50 epochs. The third one is the WeCAPP project[4] (Riffeser et al. 2003), which incorporated the 1.2 m Calar Alto telescope and the 0.8 m Wendelstein telescope, to monitor a smaller field (16 × 16 arcminutes) centered on the bulge for a decade from 1999 till 2008. While the microlensing searches and constraints on the compact dark matter fraction in the halo of M31 remain inconclusive, these data sets provide tremendous leverage to the variable populations in M31.

1.4 Wide-Field Camera View of M31

The third wave of M31 surveys comes hand-in-hand with the ultra-wide cameras. There are two M31 monitoring programs using different ultra-wide detectors. The first one is the Pan-STARRS Andromeda project (Lee et al. 2012), which utilizes the Giga-pixel camera (GPC) on-board the 1.8 m Pan-STARRS 1 telescope at Haleakala. Pan-STARRS is capable of imaging ~7 squared degrees, incorporating the bulge, disk, halo of M31, and even the companion galaxy NGC 205, in a single shot. PAndromeda observed M31 from 2010 to 2012, with ~300 epochs and intra-night cadence to hunt for microlensing events. The main goal of PAndromeda is to put a better constraint on the compact dark matter fraction in the halo of M31 with larger number of microlensing events. Nevertheless, the high-cadence data from PAndromeda also provides tremendous leverage to the studies of variables. Another ultra-wide camera project to monitor M31 is the Palomar Transient Factory (PTF). PTF made use of the refurbished CFH12K camera, mounted on the 1.2 m Schmidt telescope at the Palomar Observatory. While the original goal of PTF is to survey nearby galaxies with cadence as high as several minutes to identify transients as early as possible, it also spared some time to observe M31, with the goal to identify recurrent novae (RNe) and put a stringent constraint on RNe as progenitors of SNe Ia (Cao et al. 2012). The PTF project has been observing M31 since 2009, with varying cadences depending on the weather. Besides RNe, the PTF data are also useful for other aperiodic variables, for example R Coronae Borealis stars (Tang et al. 2013). A summary of different waves of the M31 monitoring program, as well as their spatial coverage and time-span, can be found in Table 1.1 and Figure 1.2.

1.5 Deep, Multi-Wavelength View of M31

Besides the cadenced observations of dedicated fields in the optical, there were also numerous deep observations across the spectrum, covering not only the bulge and the disk of the Andromeda galaxy, but extending to its halo and reaching out to its satellites (including the Triangulum galaxy, Messier 33). The wealth of information provided by these deep, multi-wavelength observations paints a complete picture of the Andromeda galaxy and its surrounding environment. More importantly, it is the deep observations into the halo of the Andromeda galaxy that provide us clues to how the Andromeda galaxy formed, hence making the Andromeda galaxy an ideal testbed for cosmology. We can compare the observational data of the Andromeda galaxy with the cosmological simulations to bring insight to our understanding of

[4] http://www.usm.uni-muenchen.de/people/arri/wecapp.html.

Table 1.1. Previous M31 Surveys

Name	Telescope	Pointing	CCD FoV	Duration
		First Wave: Searching for Distance Indicators		
Direct project	FLWO 1.2 m, MDM 1.3 m	Along the spiral arms	11′ × 11′	1996–2000
Ribas et al.	INT 2.5 m	Northern disk	33′ × 33′	1999–2003
		Second Wave: Hunting for Microlensing		
WeCAPP	Wendelstein 0.8 m, Calar Alto 1.23 m	M31 bulge	16′ × 16′	1999–2008
POINT-AGAPE	INT 2.5 m	Northern and southern disks	33′ × 33′	1999–2001
POMME	CFHT 3.6 m	Northern and southern disks	1 × 1 deg	2004–2005
		Third Wave:		
PAndromeda	Pan-STARRS 1.8 m	Entire M31, including NGC 205	3 deg in diameter	2010–2012
PTF/iPTF	Samuel Oschin Telescope 1.2 m	Entire M31, including NGC 205	2.33.5 deg	2009–2017

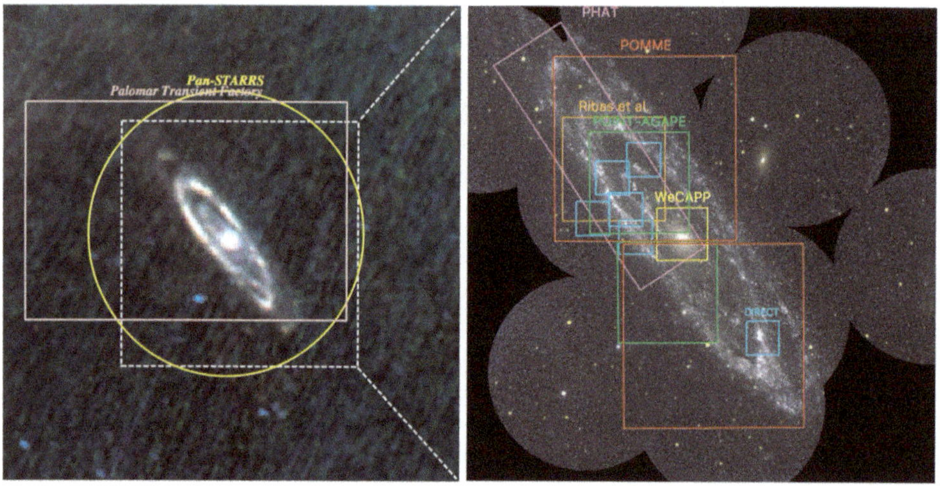

Figure 1.2. Left: Footprints of the Pan-STARRS (yellow circle) and Palomar Transient Factory (pink rectangle) coverage of M31, overlaid on a tri-color composite image from the *Infrared Astronomical Satellite* (*IRAS*). Right: Footprints of different M31 time-series surveys listed in Table 1.1. We also mark the Panchromatic Hubble Andromeda Treasury (PHAT, violet rectangle). PHAT is not a time-series survey, but provides a deep and sharp view over 1/3 of the M31 disk in UV, optical, and infrared pass-bands. The underlying image is a tri-color composite image from the *GALEX* satellite.

cosmology (this is often dubbed "near-field cosmology"). One important observational feature to compare with the cosmological simulations is the spatial distribution of the tidal debris, substructures, and satellite galaxies seen in the halo of large galaxies like the Milky Way and the Andromeda galaxy. This is because the spatial distribution of such features depends on the underlying physics and that enters into the cosmological model (which is often assumed to be ΛCDM). These features can thus be used to test the parameters in the cosmological simulations and shed light on how galaxies are assembled across cosmic times. That being said, there are several ways to identify substructures (be it tidal trails or satellite galaxies). The first method is to identify over-density of stars. This can be performed if we have a homogeneous coverage of a large portion of the sky. For example, using a simple color cut, Belokurov et al. (2006) can clearly identify the tidal streams of the Sagittarius dwarf spheroidal galaxy in the Sloan Digital Sky Survey (SDSS) imaging around the north Galactic cap. Likewise, using deep i-band and V-band imaging of the Andromeda galaxy from the Isaac Newton Telescope and the Wide Field camera, Ibata et al. (2001) were able to identify red giant branch stars (RGBs) with the color and apparent magnitude at the distance of the Andromeda galaxy. From the spatial distribution of the RGBs, they found a giant stream of metal-rich stars in the halo of the Andromeda galaxy. This giant stream is a piece of evidence that the Andromeda galaxy has tidally interacted and possibly accreted a companion galaxy (Figures 1.3 and 1.4).

In addition to counting stars and finding over-density of stars as an indication of substructures, another approach to identify substructures is via variable stars. This is

Figure 1.3. A map of stars in the outer regions of the Milky Way, derived from the SDSS images of the northern sky. The color indicates the distance of the stars, while the intensity indicates the density of stars on the sky. Structures visible in this map include streams of the stars torn from the Sagittarius dwarf spheroidal galaxy (extending from the lower left to the lower right of this figure), a smaller orphan stream crossing the Sagittarius streams, the "Monoceros Ring" that encircles the Milky Way disk, trails of stars being stripped from the globular cluster Palomar 5, and excesses of stars found toward the constellations Virgo and Hercules. Image credit: Belokurov and the SDSS (https://www.sdss.org/science/idl-tiff-file-2/), under standard SDSS image license.

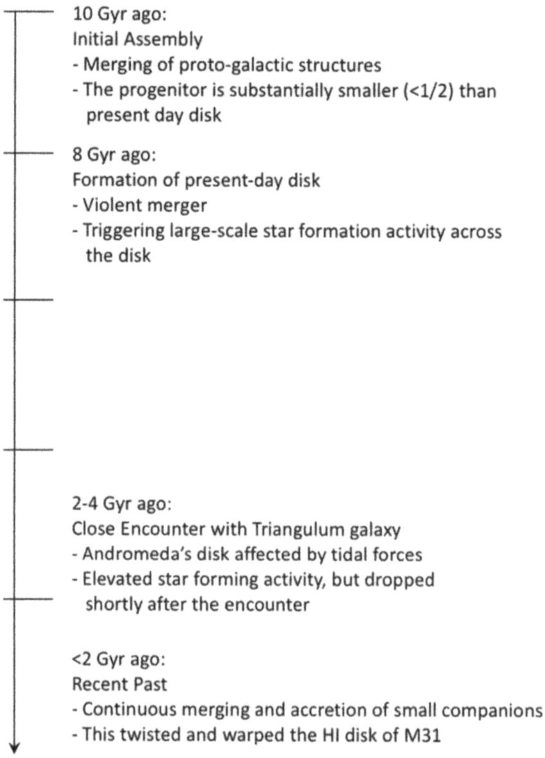

Figure 1.4. A concise summary of the formation and assembly of the disk of the Andromeda galaxy, extracted from Davidge et al. (2012).

especially the case as tidal streams that are seen nowadays would have originated from interactions of galaxies billions of years ago. This means that we can clearly see that the tidal features from variable stars stem from old stellar population, e.g., RR Lyra stars that are at an age of >10 Gyr. For example, Torrelaba et al. (2015) have shown that from the Galactic RR Lyra stars cataloged by the Catalina Sky Survey, we can clearly identify the Sagittarius streams across the sky. However, we should note that RR Lyra stars are rather faint at the distance of the Andromeda galaxy. This means that we will need to resort to large ground-based telescopes with 8 m aperture, or use the *Hubble Space Telescope*, to identify RR Lyra stars at the distance of the Andromeda galaxy (see discussion in Chapter 6 of this book).

With observational constraints, and in the context of near-field cosmology, Davidge et al. (2012) have gathered a picture of the formation and assembly of the disk of the Andromeda galaxy. Here we provide a concise summary of the time lines outlined by Davidge et al. (2012), as they are pertinent to understanding the distribution of different types of variables that we see in the Andromeda galaxy. In addition, we can also use properties of the variables (e.g., spatial distribution, age, metallicity, etc.) to locate uncharted tidal trails and to provide tests on the formation and assembly process of the Andromeda galaxy. A thorough review of multi-wavelength studies of the Andromeda galaxy is provided in Chapter 5 of this book.

References

Aurière, M., Baillon, P., Bouquet, A., et al. 2001, ApJ, 553, L137
Baade, W. 1944, ApJ, 100, 137
Baade, W. 1956, PASP, 68, 5
Belokurov, V., Zucker, D. B., Evans, N. W., et al. 2006, ApJ, 642, L137
Cao, Y., Kasliwal, M. M., Neill, J. D., et al. 2012, ApJ, 752, 133
Crotts, A. P. S., & Tomaney, A. B. 1996, ApJ, 473, L87
Davidge, T. J., McConnachie, A. W., Fardal, M. A., et al. 2012, ApJ, 751, 74
de Vaucouleurs, G., de Vaucouleurs, A., Corwin, H. G. Jr., et al. 1991, Third Reference Catalogue of Bright Galaxies. Volume I: Explanations and references. Volume II: Data for galaxies between 0^h and 12^h. Volume III: Data for galaxies between 12^h and 24^h (New York: Springer)
Einstein, A. 1936, Science, 84, 506
Hartwig, E. 1885, AN, 112, 355
Hubble, E. P. 1926, ApJ, 64, 321
Hubble, E. P. 1929, ApJ, 69, 103
Ibata, R., Irwin, M., Lewis, G., Ferguson, A. M. N., & Tanvir, N. 2001, Nature, 412, 49
Lee, C.-H., Riffeser, A., Koppenhoefer, J., et al. 2012, AJ, 143, 89
Macri, L. M. 2004, in ASP Conf. Ser. 310, IAU Coll. 193: Variable Stars in the Local Group, ed. D. W. Kurtz, & K. R. Pollard (San Francisco, CA: ASP), 33
Paczynski, B. 1986, ApJ, 304, 1
Ribas, I., Jordi, C., Vilardell, F., Giménez, Á., & Guinan, E. F. 2004, NewAR, 48, 755
Riess, A. G., Fliri, J., & Valls-Gabaud, D. 2012, ApJ, 745, 156
Riess, A. G., Macri, L. M., Hoffmann, S. L., et al. 2016, ApJ, 826, 56
Riffeser, A., Fliri, J., Bender, R., Seitz, S., & Gössl, C. A. 2003, ApJ, 599, L17
Rubin, V. C., & Ford, W. K. Jr. 1970, ApJ, 159, 379
Scheiner, J. 1899, ApJ, 9
Slipher, V. M. 1913, LowOB, 2, 56
Tang, S., Cao, Y., Bildsten, L., et al. 2013, ApJ, 767, L23
Torrealba, G., Catelan, M., Drake, A. J., et al. 2015, MNRAS, 446, 2251
Vilardell, F., Ribas, I., Jordi, C., Fitzpatrick, E. L., & Guinan, E. F. 2010, A&A, 509, A70
Zwicky, F. 1933, AcHPh, 6, 110

AAS | IOP Astronomy

Time-Domain Studies of the Andromeda Galaxy

Chien-Hsiu Lee

Chapter 2

Techniques for Time-Domain Studies

2.1 Image Subtraction

Finding varying sources, especially in dense stellar fields like the Milky Way bulge or the Andromeda galaxy, is not an easy task. This is because, due to the crowding and blending effects from nearby sources, it is difficult to extract photometry and light curves accurately even with PSF photometry, let alone differential photometry. To tackle these issues, there have been suggestions of conducting image subtraction (see Figure 2.1) that allows not only identifying varying sources, but also extracting high quality light curves. The idea of image subtraction is that we match two images to the same flux level and to the same PSF, such that after subtraction, the varying sources will stand out. The most critical step of image subtraction is to take into account the seeing variation among images and to find an appropriate kernel that can homogenize the PSF among images:

$$\mathrm{Ref}(x,\, y) \otimes \mathrm{Kernel}(u,\, v) = I(x,\, y), \tag{2.1}$$

where $\mathrm{Ref}(x, y)$ and $I(x, y)$ denote the reference and individual images, respectively. The convolution $\otimes$ is carried out using $\mathrm{Kernel}(u, v)$. The main hurdle to derive $\mathrm{Kernel}(u, v)$ is that it is a non-linear process to solve this equation. This will be computationally expensive when it comes to finding the kernels for hundreds to thousands of images.

Tomaney & Crotts (1996) first proposed to adopt the Fourier transform ratio of bright stars in different images to construct the kernel. However, as the Fourier transform at high frequency is dominated by noises, they found that this method hardly worked.

Kochanski et al. (1996) tackled this issue from another angle. They proposed to find the kernel by minimizing the difference between two images. In this approach, however, finding the kernel is a non-linear least-square fitting process. Such a process requires vast amounts of computation time, making it impossible to determine the kernel even in uncrowded fields where only a small fraction of the

doi:10.1088/2514-3433/ab3639ch2

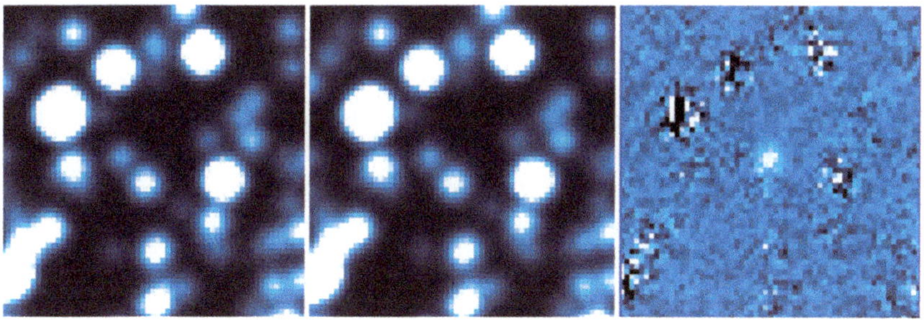

Figure 2.1. Example of image subtraction. When subtracting the first epoch (left) from the second epoch (middle), the variable stands out (right). Data taken from the ISIS package by Alard (2000).

pixels is used to sample the PSF. To get around with the non-linear process, Alard & Lupton (1998) proposed to approximate the Kernel with the summation of a series of Gaussian functions and polynomials, hence making this a linear least-square fitting process

$$\text{Kernel}(u, v) = \sum_{n}\sum_{d_n^x}\sum_{d_n^y} a_n \times e^{-(u^2+v^2)/2\sigma_n^2} u^{d_n^x} v^{d_n^y}, \tag{2.2}$$

where $0 < d_n^x \leqslant D_n$, $0 < d_n^y + d_n^x \leqslant D_n$, and D_n is the degree of the polynomial corresponding to the nth Gaussian component.

Through this approximation, they found that using three Gaussian components with polynomials of degrees between two and six can yield decent image subtraction for most of the cases. They applied this method to very dense stellar field images taken from the OGLE microlensing survey, and demonstrated that the subtracted images can reach residuals down to the photon noise level. Another advantage of the Alard & Lupton (1998) procedure is that the background level in different images can also be fitted simultaneously during the linear least-square fitting process as

$$\text{Ref}(x, y) \otimes \text{Kernel}(x, y) = I(x, y) + \text{Bg}(x, y), \tag{2.3}$$

where

$$\text{Bg}(x, y) = \sum_{i}\sum_{j} a_i x^i y^j, \tag{2.4}$$

with $0 < i \leqslant D^{\text{Bg}}$, $0 < i + j \leqslant D^{\text{Bg}}$, and D^{Bg} is the degree of the polynomial to take into account the background difference.

Since this improvement, the image subtraction procedure has been implemented by numerous researchers, e.g., Alard (2000),[1] Wozniak (2000),[2] Bond (2001), Goessl & Riffeser (2002),[3] to name a few; the later three implementations are especially tailored for OGLE, MOA, and WeCAPP microlensing surveys of dense stellar fields in the Galactic bulge, Magellanic Clouds, and the Andromeda galaxy.

[1] The ISIS package, http://www.iap.fr/users/alard/package.html.

[2] The DIAPL package, http://users.camk.edu.pl/pych/DIAPL/.

[3] The mupipe package, http://www.usm.uni-muenchen.de/people/arri/mupipe/.

2.2 Period Search Algorithms

After identifying the varying sources, the next step is to see if the varying sources are periodic. If they are periodic, we also want to determine their periods (see, e.g., Figures 2.2–2.4). There are several ways to determine the period. The most widely employed method is the Lomb–Scargle (LS) algorithm (Lomb 1976; Scargle 1982), which uses a Fourier transform to calculate the power of a given frequency (1/*period*) in the light curves. While widely used, there is a caveat that instead of the true period, the LS method might return the observation cadence (and alias) as the highest power frequency. This is especially the case for repeated observations with daily cadences.

To better avoid this misidentification, there are also other period estimate algorithms that do not use the Fourier transform method. For example, the analysis of variance method—proposed by Schwarzenberg-Czerny (1989)—folds a light curve with different periods and computes the variances of the phased light curve. At the correct period, the data points should align with each other at a given phase, hence we can determine the period by minimizing the variance of the phased light curve. If the variable shows periodic dimming, e.g., eclipsing binaries or transiting exoplanets, the box-fitting least-square method by Kovács et al. (2002) may be the most ideal algorithm to apply. The box-fitting algorithm looks for periodic alternation between two discrete levels, with much less time spent at the lower level. The advantage of the box-fitting algorithm is finding very small dimming signals with low S/N (or large photometric uncertainties), hence it is especially suitable for exoplanet transits, whose dimming signals are at the level of 1% or less.

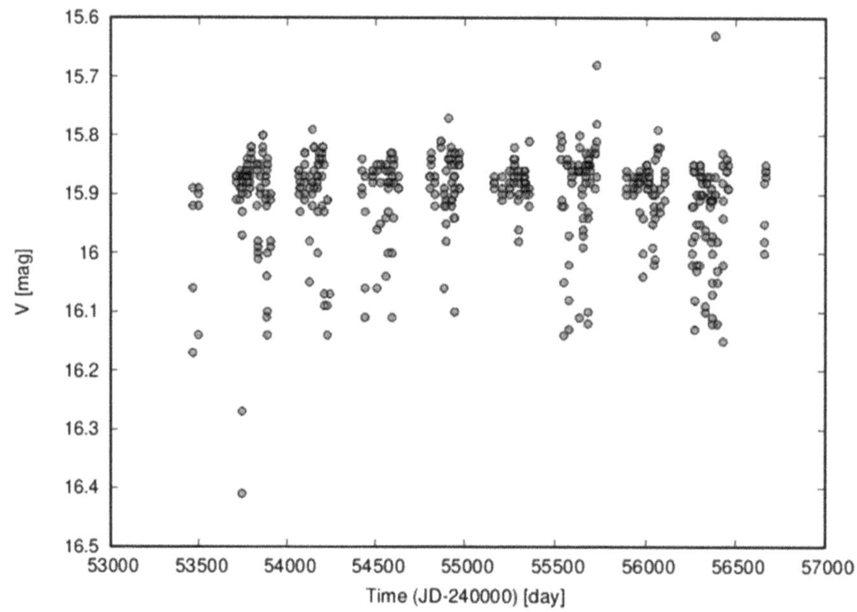

Figure 2.2. Example of an unfolded light curve of an eclipsing binary. Data from Catalina Sky Survey.

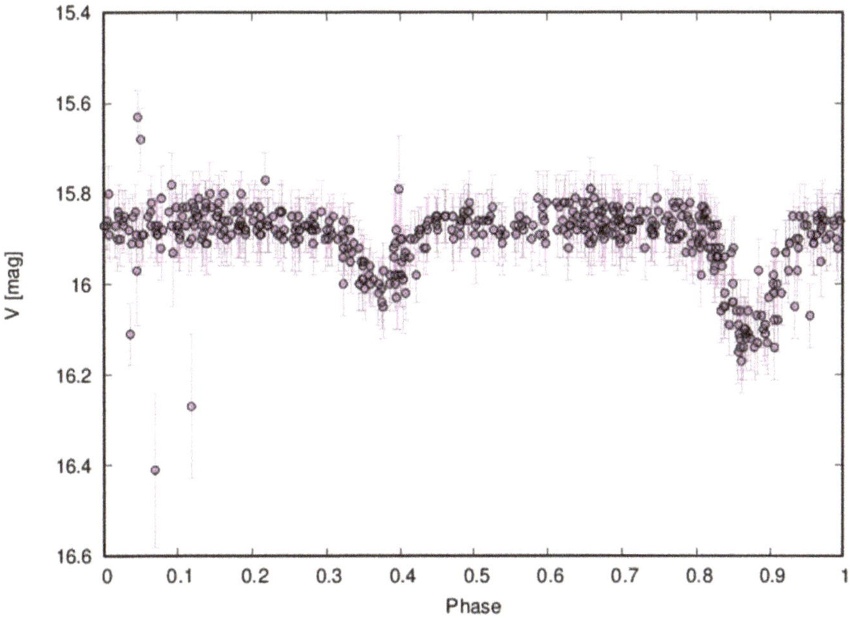

Figure 2.3. Example of a folded light curve of an eclipsing binary. This is folding the light curve of Figure 2.2 with the correct period (as shown in Figure 2.4). Data from Catalina Sky Survey.

For very poorly sampled light curves (with only 10–20 data points) where the aforementioned methods fail, we can also use the template fitting method by Layden & Sarajedini (2000). For some classes of variables, e.g., RR Lyra stars, the light curves all have similar shapes and we can construct a characteristic template light curve from a large sample. We can then fit the poorly sampled light curves with this template and determine the period. The beauty of template fitting is that we can also obtain some key features of each individual variable, e.g., mean magnitude, magnitude maximum and minimum, rise time, to name a few. This is especially important to derive the period–luminosity relation of RR Lyra stars and Cepheids and use them as distance indicators in very distant galaxies. At large distance where the variables are faint, we need relatively long exposure time to obtain decent photometry and it is expensive to obtain many epochs. For example, Hoffmann et al. (2016) used the template-fitting method to obtain the mean magnitudes and period–luminosity relation of Cepheids in 19 distant, Type Ia SNe host galaxies.

Most of the aforementioned algorithms (and many more) are implemented by Hartman in his publicly available VARTOOLS program (Hartman & Bakos 2016). In addition, the NASA Exoplanet Archive also provides online periodogram tools,[4] which can be handy for a quick search for periods with some selected algorithms. The Template-Fitting method has been implemented by Andrew Layden and is also publicly available,[5] which includes 10 templates for different types (RR Lyra stars,

[4] https://exoplanetarchive.ipac.caltech.edu/cgi-bin/Pgram/nph-pgram.

[5] http://physics.bgsu.edu/layden/ASTRO/DATA/EXPORT/progs.htm.

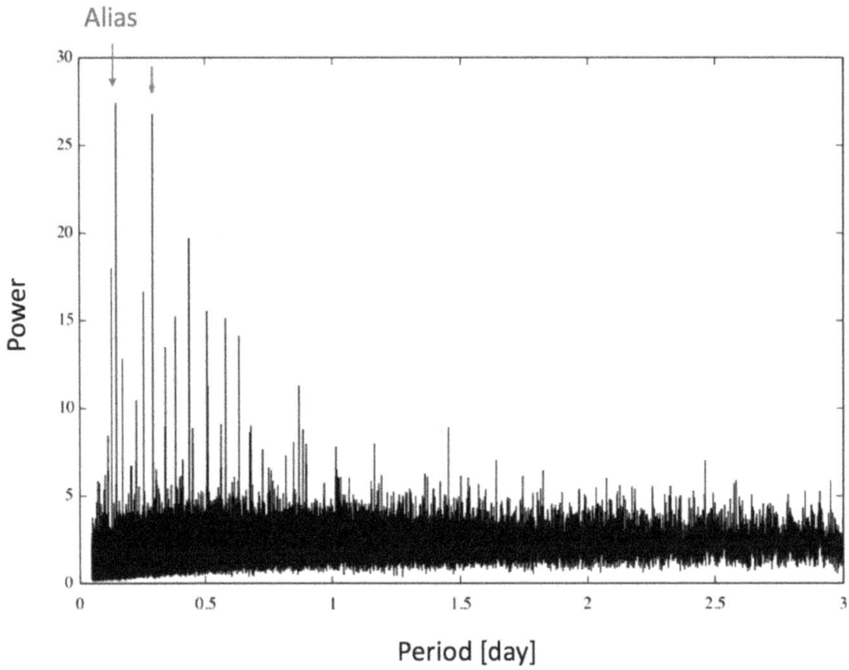

Figure 2.4. Periodogram of the eclipsing binary light curve from Figure 2.2. The true period (0.29 days) is marked by the red arrow. Please note that there are two periods that have power greater than 25. The other period ~0.145 days is an alias of the true period of 0.29 days. This is common in eclipsing binary period search, especially when the primary and secondary eclipse depths are similar, the period search program is likely to return half of the true period.

Cepheids, eclipsing binaries) of variables. For Cepheids in particular, there have been extensive light curve templates in different filters, e.g., in B-, V-, and I-band by Peter Yoachim[6] (Yoachim et al. 2009). Ondrej Pejcha (Pejcha & Kochanek 2012) provided a wider variety of filters from optical to near-infrared and even predicted light curves in new filters.[7]

2.3 Classification

After identifying the variability, the next step is to determine the type of varying sources. Variables can be roughly fit into three flavors: eruptive, pulsating, and eclipsing. A simple way to distinguish among these three flavors is to use the ratio of the number of light curve points above and below the median magnitudes. On the one hand, for eruptive variables, most of the data points will lie on the baseline where the median magnitude is, and we will see a large above versus below ratio, where nearly no data points are below the median. For eclipsing variables, on the other hand, we expect to see a very small above/below ratio. As for pulsating

[6] http://staff.washington.edu/yoachim/code.php.
[7] https://www.astro.princeton.edu/pejcha/cepheids/.

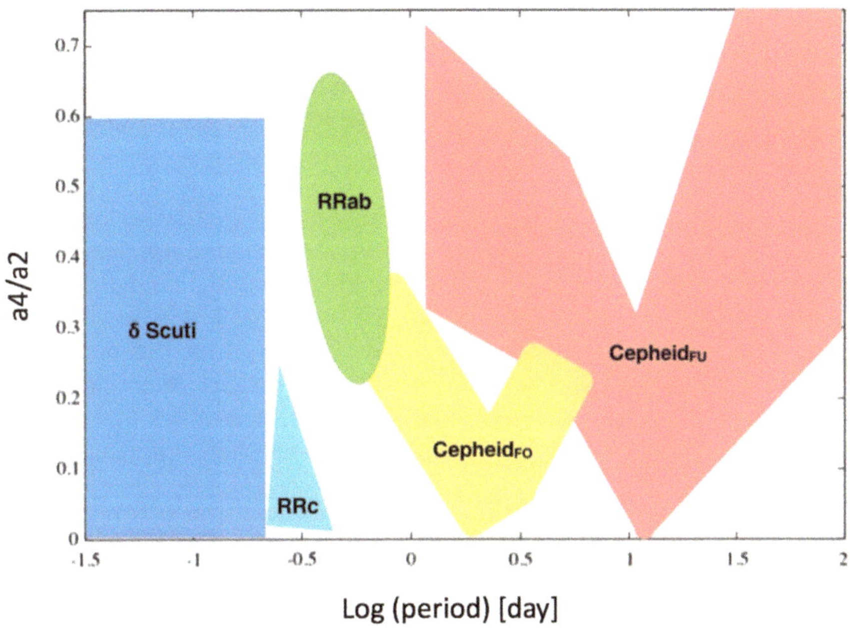

Figure 2.5. Classification of pulsating variables using log(periods) and a_4/a_2.

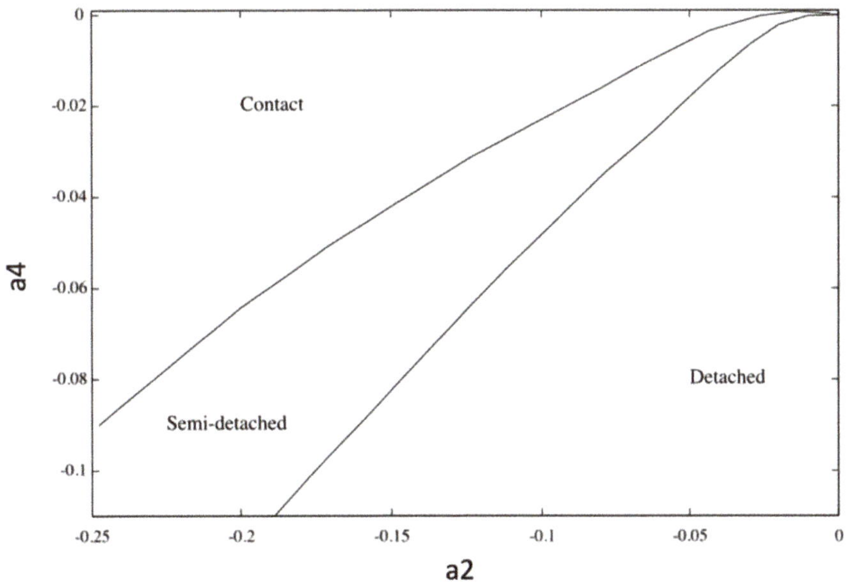

Figure 2.6. Classification of eclipsing binaries using a_2 versus a_4.

variables, the light curves will be symmetric above and below the median magnitudes, and the above/below ratio is expected to be close to unity.

After determining the flavor of the variables, we can proceed to more sophisticated classifications. The pulsating variables, for example, are composed of many different classes, including RR Lyra a/b stars, RR Lyra c stars, Cepheids in fundamental mode, Cepheids in first overtone, Delta Scuti stars, to name a few. These different classes of variables follow different period–luminosity relations, so if we know the distance or can determine their absolute magnitudes, then we can use the period–luminosity relation to classify them. However, obtaining distance can be a difficult task, either in extra-galactic systems or even within the Milky Way. Furthermore, as we use these variables as distance indicators, this means that we often do not have their distance a priori. Another way to determine their classes is via Fourier decomposition. The idea is that we can reconstruct the phased light curves (or some of the main features) using a series of sine and cosine functions in the form of:

$$f(\phi) = \sum_{i=1}^{n} a_i \cos(2\pi i\phi) + b_i \sin(2\pi i\phi), \tag{2.5}$$

where $f(\phi)$ is the normalized flux at phase ϕ. This approach was first envisioned by Schaltenbrand & Tammann (1971) and later on further developed by Simon & Lee (1981) to understand the structural properties of light curves of Cepheids. While there has not been a consensus of which parameters correspond to the distinct features of different variables, this approach has been used empirically to connect the parameters of the Fourier decomposition to different types of variables, with the aim to categorize them based on the Fourier decomposition parameters as well as the periods. For example, Grzegorz Pojmanski (Pojmanski 2002) has shown that, using the Fourier decomposition coefficients a_4/a_2 and the periods, we can efficiently classify the pulsating variables, as illustrated in Figure 2.5. Grzegorz Pojmanski (2002) also compared simulated eclipsing binary light curves from the state-of-the-art binary light curve modeling by Wilson & Devinney (1971). He found that for the eclipsing binaries, we can also use a similar trick, with merely a_2 and a_4, to classify eclipsing binaries into classes of detached, semi-detached, or over-contact. These classes correspond to their Roche-lobe filling configuration (see Figure 2.6). The detached eclipsing binaries are of special interests because both of the stars in the binary system are not filling their Roche-lobes, so we can use them to determine the fundamental properties of stars. The over-contact binaries, on the other hand, follow an empirical period–luminosity relation and are handy for distance determination (though with larger uncertainties when compared to Cepheids or RR Lyra stars). However, we should note that the Fourier decomposition and the derived coefficients alone may not be sufficient to accurately classify variables. In this regard, we can extract a wide array of light curve features (on top of the Fourier coefficients), and use machine learning to sort out a light curve with probability of belonging to a given class (see, e.g., Richards et al. 2012).

References

Alard, C. 2000, A&AS, 144, 363
Alard, C., & Lupton, R. H. 1998, ApJ, 503, 325
Bond, I. 2001, in ASP Conf. Proc. 239, Microlensing 2000: A New Era of Microlensing Astrophysics, ed. J. W. Menzies, & P. D. Sackett (San Francisco, CA: ASP), 33
Gössl, C. A., & Riffeser, A. 2002, A&A, 381, 1095
Hartman, J. D., & Bakos, G. Á. 2016, A&C, 17, 1
Hoffmann, S. L., Macri, L. M., Riess, A. G., et al. 2016, ApJ, 830, 10
Kochanski, G. P., Tyson, J. A., & Fischer, P. 1996, AJ, 111, 1444
Kovács, G., Zucker, S., & Mazeh, T. 2002, A&A, 391, 369
Layden, A. C., & Sarajedini, A. 2000, AJ, 119, 1760
Lomb, N. R. 1976, Ap&SS, 39, 447
Pejcha, O., & Kochanek, C. S. 2012, ApJ, 748, 107
Pojmanski, G. 2002, AcA, 52, 397
Richards, J. W., Starr, D. L., Miller, A. A., et al. 2012, ApJS, 203, 32
Scargle, J. D. 1982, ApJ, 263, 835
Schaltenbrand, R., & Tammann, G. A. 1971, AAPS, 4, 265
Schwarzenberg-Czerny, A. 1989, MNRAS, 241, 153
Simon, N. R., & Lee, A. S. 1981, APJ, 248, 291
Tomaney, A. B., & Crotts, A. P. S. 1996, AJ, 112, 2872
Wilson, R. E., & Devinney, E. J. 1971, APJ, 166, 605
Wozniak, P. R. 2000, AcA, 50, 421
Yoachim, P., McCommas, L. P., Dalcanton, J. J., & Williams, B. F. 2009, AJ, 137, 4697

Chien-Hsiu Lee

Chapter 3

Periodic and Aperiodic Variables

3.1 Periodic Variables

Periodic variables come in different flavors, here we focus on two of them, i.e., pulsating and eclipsing. The physics behind pulsating variables is the tug-of-war between thermal pressure and gravity. This idea was proposed by Sir Arthur Eddington to explain the pulsation of Cepheids, the so-called Eddington's valve (Eddington 1917). The idea of Eddington's valve is that when the stellar photosphere expands, the temperature decreases, such that the gas in the photosphere is deionized and the opacity decreases and cannot capture the stellar radiation effectively. This decreases the expansion, until it completely stops, and then begins contraction due to gravity. When it contracts, the temperature increases, leading to ionization of the gas. This allows more efficient capture of stellar radiation and at some point reverses the contraction and starts another cycle of expansion. The acting agent (or type of gas) was identified by Zhevakin (1963) as ionized helium. Another important factor of stellar pulsation depends on where the partial ionization zone is. At temperatures greater than 7500 K, the partial ionization zone would be too close to the stellar surface, hence there is not enough mass to drive the oscillation. On the other hand, at temperatures less than 5500 K, the partial ionization zone would be too deep in the stellar interior where the convection will prevent thermal pressure accumulating. The variable stars can be identified on the HR diagram in the instability strip, bracketed by the temperature limits mentioned above. The instability strip intersects with the main sequence; the intersection is where the δ Scuti stars reside. Evolved stars also enter the instability strip. For example, Cepheids are evolved supergiants that come across the instability strip. The RR Lyra stars, on the other hand, are evolved stars at the intersection between the horizontal branch and instability strip. In this chapter, we focus mainly on the Cepheids, as their luminosity renders them within reach of ground-based observations at the distance of M31. The RR Lyra stars, on the other hand, are too dim to be identified by mid-class telescopes at the distance of M31.

doi:10.1088/2514-3433/ab3639ch3

The other kind of variables that we focus on is eclipsing binaries. Eclipsing binaries exhibit brightness variation due to the binary stars occulting each other, and cause dimming of the ensemble brightness. As mentioned in Chapter 2, eclipsing binaries can be categorized into three classes depending on their Roche-lobe filling configuration. Originally it was thought that detached eclipsing binaries are the best targets for distance estimates, because there is no Roche-lobe filling of both stars. This makes the modeling and interpretation relatively easier than with semi-detached and over-contact binaries. However, as have been shown in Wyithe & Wilson (2002), modeling detached eclipsing binary light curves has the disadvantage of degeneracy in the mass ratio. The semi-detached eclipsing binaries, on the other hand, do provide a more stringent constraint of the mass ratio thanks to the fact that one of the stellar components is filling its Roche-lobe. This is great leverage in increasing the number of eclipsing binaries for distance estimates, especially using eclipsing binaries to determine extra-galactic distances like to the Andromeda galaxy, as there are only a handful of eclipsing binaries that are bright enough to obtain multi-epoch spectra, which still requires 8–10 m telescopes with hours of observations. As a matter of fact, independent M31 distance estimates from eclipsing binaries were obtained via two semi-detached eclipsing binary systems (Ribas et al. 2005; Vilardell et al. 2010). It has been demonstrated in these previous works that using semi-detached eclipsing binaries can reduce the number of free parameters in modeling the binary light curves and spectra, albeit at the cost of having to account for other effects. Nevertheless, if these effects are properly considered, there should not be systematic error in the derived distances. We should also note that the state-of-the-art modeling effort, especially the Wilson-Devinney code, can also accurately model the over-contact binaries. In addition, there has also been period–luminosity relations for over-contact binaries from photometric measurements alone, reaching a precision of 8% (Chen et al. 2018).

In this chapter, we will further explore specific types of pulsating variables, i.e., Cepheids, as well as eclipsing binaries, and show how to use them not only to infer the distance, but for other properties that can be used to study stellar astrophysics and M31 per se.

3.1.1 Cepheids as Distance Anchors

Cepheids are variables that pulsate periodically due to the change of their radius and effective temperature. Leavitt was the first to recognize a period–luminosity relation of Cepheids when examining and recording the brightness variation of variables in the Large Magellanic Clouds using photometric plates (Leavitt & Pickering 1912). The Leavitt lLaw makes Cepheids ideal distance indicators, reaching to distances as far as tens of Mpc. The Leavitt lLaw paved the way to determine the distance of extra-galactic objects, with which Edwin Hubble was able to estimate the distance of M31 and proved it to be an island universe, thus settling the 1920 Great Debate between Harlow Shapley and Heber Curtis. Even more so, Cepheids have been an indispensable rung on the cosmic distance ladder. For example, the key project of the *Hubble Space Telescope* is to resolve the discrepancies of Hubble constant estimates, ranging from 50 km s^{-1} Mpc^{-1} to 100 km s^{-1} Mpc^{-1} in the 1970s. The final results from the *HST* Key project delivered a 10% Hubble constant of $H_0 = 72 \pm 8$ km s^{-1} Mpc^{-1} (Freedman et al. 2001). Recently, joining the forces of Cepheids and SNe Ia has led to the discovery of dark energy and the accelerating Universe (Riess et al. 1998; Perlmutter et al. 1999). The current use of Leavitt's Law for cosmology studies is heavily relying on three local distance anchors that we can determine the distance geometrically and precisely. The first one is our Milky Way, where we have a dozen Cepheids with accurate distance estimates from the parallax method (Riess et al. 2018). The second one is the Large Magellanic Cloud, where Pietrzynski et al. (2013) have determined its distance to a 2% level using eclipsing binaries from late-type stars and the empirical color-angular size relation from interferometry. The third one is the maser galaxy NGC 4258; with the high spatial resolution power delivered by VLBI, we are able to measure the orbital motion of the masers in the circum-nuclear disk, providing a geometric distance estimate of 3% (Humphreys et al. 2013). However, each of the distance anchors, while having the distance estimate as precise as 2%–3%, bear some caveats. In the Milky Way, we only have a small sample of Cepheids with known distances. Worse of all, these Cepheids reside on the short-period end, with only one Cepheid having a period of >10 days, not overlapping with the majority of longer period ($10 < P < 60$ days) Cepheids used in the cosmic distance ladder. The Large Magellanic Cloud has numerous Cepheids identified by the OGLE survey, and the well-calibrated infrared Leavitt Law thanks to the SIRIUS survey at the Infrared Survey Facility at SAAO (Ita et al. 2004). However, the Large Magellanic Cloud has relatively low metallicity compared to the distant SNe Ia host galaxies that are usually spiral galaxies, and the metallicity influence on the Leavitt Law is not well characterized. As for the maser galaxy NGC 4258, it has the largest distance uncertainty compared to the Milky Way and Large Magellanic Cloud. Furthermore, as pointed out by George Efstathiou (2014), the maser distance is in tension with its Cepheid distance if we assume a Milky Way Leavitt Law. More importantly, in the most recent Hubble constant estimate by the SH0ES team (Riess et al. 2016), it is clear the uncertainty of the distance anchor has one of the leading roles in the error budget of H_0. One possible solution is to increase the number of distance anchors, and reduce the uncertainties statistically. In this regard, M31, our closest spiral galaxy, is an ideal target to become the next distance anchor. This is especially the case because we can resolve the stellar populations at the bright end thanks to the proximity of M31. Even more so, we can tackle the crowding effect (which influences the accuracy of the Leavitt Law) given the exquisite spatial resolution from the *Hubble Space Telescope* (see Chapter 5). In addition, M31 has similar metallicity as distant SNe Ia host galaxies, hence we can investigate the metallicity influence on the Leavitt Law by comparing the Cepheids in M31 and in the Large Magellanic Cloud.

Thanks to the advent of ultra-wide camera, there have been several time-domain surveys targeting M31 with thousands of Cepheids identified. For the time being, the largest public M31 Cepheid sample is composed of ~2000 Cepheids (Kodric et al. 2013), where the light curves from the PAndromeda survey in r- and i-band are public via VizieR. There has been a larger Cepheid sample (~2500) by the POMME survey, but the data are not public yet.

3.1.1.1 Period–Luminosity Relation

Vilardell et al. (2007) and Kodric et al. (2013) have presented the Leavitt Law in M31 (see, e.g., Figure 3.1) based on ground-based photometry in B/V and r/i-bands, respectively. As can be seen from their results, the ground-based photometry suffers from crowding/blending issues due to the mediocre seeing condition. In this regard, the exquisite spatial resolution from space telescopes like the *Hubble Space Telescope* provides a unique opportunity to tackle the crowding/blending issues. Furthermore, as the light curve semi-amplitude becomes smaller with increasing wavelength, the light curve amplitude variation is negligible in the near-infrared, allowing us to establish a very tight Leavitt Law even with photometry at random phase. Another merit of the infrared observation is that the extinction effect is much smaller compared to that in the optical (six times smaller). Riess et al. (2012) first piloted the study of the M31 Cepheids in the infrared by combining Cepheids period from the ground, and exquisite infrared photometry from WFC3/*HST*. With the large Cepheid sample from PAndromeda, and the full infrared coverage of the Panchromatic Hubble Andromeda Treasury survey with WFC3/*HST*, Kodric et al. (2015) were able to not only derive a very tight and clean Leavitt Law, but also investigated and identified a broken Leavitt Law at period around 10 days.

One major concern of the period–luminosity relation (PLR) is the metallicity effect. This is especially the case because one of the local distance anchors is the Large Magellanic Cloud, which is a dwarf galaxy and comparatively metal-poor compared to other distance anchors and the more distant type Ia SNe hosts that are

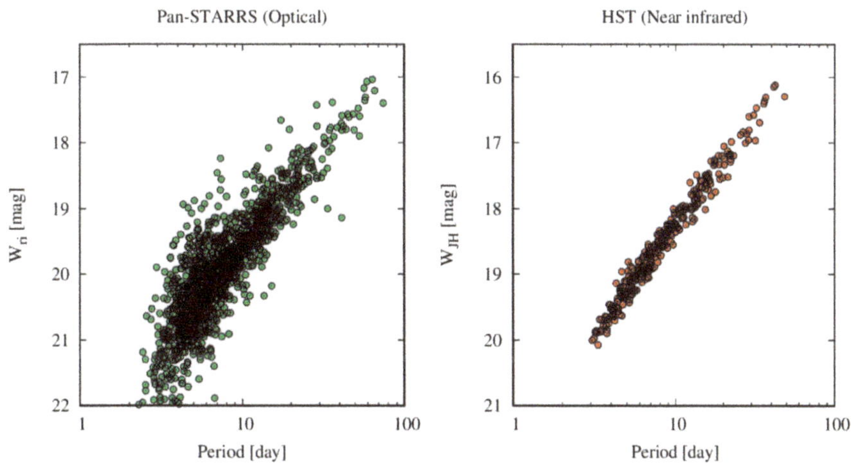

Figure 3.1. Period–luminosity relation of M31 fundamental mode Cepheids from ground-based optical (left panel) and space-based near-infrared (right panel) observations. Data taken from Kodric et al. (2013, 2015). The magnitudes are extinction-free Wesenheit magnitudes, derived from r- and i-band as $W_{ri} = r - 3.86 \times (r - i)$, as well as from HST F110W and F160W as $W_{JH} = \text{F110W} - 2.39 \times (\text{F110W} - \text{F160W})$.

mostly spiral galaxy and relatively metal-rich. In this regard, there have been some efforts to constrain the metallicity effects. For example, Freedman & Madore (2011) used a small sample of Cehepids that have well-established distance and multi-wavelength light curves to investigate how the period–luminosity relation depends on the metallicity, and their results show that the metallicity effects is insignificant in the infrared pass-bands. In addition, Majaess et al. (2011) also indicated that the V-I Wesenheit PLR is also insensitive to metallicity. The current consensus is that while there is a metallicity effect on period–luminosity relation, we can minimize or get rid of it by using either NIR or the V–I Wesenheit relation.

3.1.1.2 Ultra-Long Period Cepheids

Traditionally when probing the Leavitt Law, it is conventional to clip out Cepheids with periods shorter than 10 days or longer than 60 days. Clipping at the short period end alleviates the crowding/blending issue. The longer period end is because it has been shown that there is a flattening of the Leavitt Law or dimming of Cepheids when it comes to longer periods. Nevertheless, Cepheids with periods longer than 80 days, i.e., ultra-long period Cepheids (ULPCs), are still of great interests. This is because they have very high intrinsic luminosity ($M_I = -7.9$ mag) and can serve as distance indicators up to 100 Mpc with current and future facilities. This is especially the case in the *James Webb Space Telescope* (*JWST*) era, where ULPCs can provide a single-step distance estimate out to distant galaxies in the Hubble flow.

However, ULPCs are very rare, and most of them were found in metal-poor environments, i.e., dwarf or irregular galaxies (Bird et al. 2009). Although recent studies by Fiorentino et al. (2012) presented 37 ULPCs in 10 galaxies with a wide array of metallicity [$7.2 < 12 + \log(O/H) < 9.2$ dex]. But these 10 galaxies do not have precise distance estimates. In this regard, M31—with a distance uncertainty of 4%—turns out to be an ideal testbed for ULPCs and their Leavitt Law. As a start, Ngeow et al. (2015) made use of the long-term M31 monitoring campaign by the Palomar Transient Factory and identified 2 ULPCs in M31. By combining archival and ongoing M31 time-series data, we will be discovering more M31 ULPCs.

3.1.1.3 Period–Age Relation

Besides the Leavitt Law and distance determination, Cepheids can also serve as age indicators. This is because if we assume Cepheids follow the Leavitt Law, the mass–luminosity relation, and the ages predicted by stellar evolutionary models, then the longer the period, the higher the mass of a Cepheid. From the stellar evolutionary model, the higher the mass, the younger the age, hence we can use the period to estimate the age of a Cepheid. Indeed, there is a well-established period–age relation by Bono et al. (2005) for classical Cepheids. Since classical Cepheids are from Population I stars, we can use them to trace recent star formation history. Such unique properties of Cepheids have been exploited by Kodric et al. (2013) using the PAndromeda Cepheid sample, where they found that classical M31 Cepheids align with the 10 kpc star-forming ring reported by Gordon et al. (2006) and Davidge et al. (2012). Even more so, using the period–age relation, we can see an age gradient across the 10-kpc ring, where the farther away from the ring, the older the Cepheids are. This result is contradicting previous suggestions that this ring-like structure is formed from M31–M32 interaction. Because in such a galaxy interaction scenario, the ring structure should propagate outward, meaning the farther away from the ring, the younger the Cepheids are. In addition, recent observations by the Panchromatic Hubble Andromeda Treasury indicates that the ring is stationary and not expanding outwards, which also poses challenges to the M31–M32 interaction scenario.

3.1.1.4 Metallicity Indicator

While Cepheids mostly pulsate in a single period, there is a rare class of beat Cepheids that exhibit two periods simultaneously due to two radial pulsation modes. Pieter Theodorus Oosterhoff (1957a, 1957b) first pioneered the work of beat Cepheids, where he employed two periods to explain the large scatters in the light curves of U TrA and TU Cas. While there were several attempts to search for further beat Cepheids in the Milky Way (Pike & Andrews 1979; Henden 1979, 1980), only two-dozen Galactic beat Cepheids have been known so far. Aside from our Milky Way, there have been extra-galactic samples of beat Cepheids thanks to microlensing surveys. The MACHO team first reported 45 beat Cepheids in the Large Magellanic Cloud (Alcock et al. 1995). The OGLE team also reported 76 beat Cepheids in the Large Magellanic Cloud (Soszynski et al. 2000) and 93 beat Cepheids in the Small Magellanic Cloud (Udalski et al. 1999). The EROS team also recently reported more than 200 beat Cepheids in the Large and Small Magellanic Clouds (Marquette et al. 2009). With the continuing observation of OGLE-III, they reported the largest sample of beat Cepheids in both the Large and Small Magellanic Clouds, with 271 in the former and 277 beat Cepheids in the later, respectively. As OGLE has started its fourth phase (OGLE-IV), these numbers will increase in the future.

One important application of beat Cepheids is they can be used as metallicity tracers (see Figure 3.2). This is because from the pulsating model's point of view, there are only four key parameters, i.e., mass, luminosity, effective temperature, and metallicity. The uniqueness of beat Cepheids is they have two pulsating periods, which gives us two constraints to the pulsating model. Then we also have another constraint from the mass–luminosity relation. The last constraint comes from the fact that Cepheids are evolved stars crossing the instability strip, where there is only a very small allowed sub-region in the effective temperature. Combining these four constraints, we can narrow down the possible metallicity niche of beat Cepheids, thus deriving the metallicity of the local environments. The drawback of using beat Cepheids is that their periods are very short, meaning they are at the faint end of the Leavitt Law and requires dedicated monitoring campaigns with decent depth and cadence to discover. This is especially the case at the distance of M31 and M33. Indeed, Beaulieu et al. 2006 have used the deep images provided by the 3.6 m CFHT M33 survey (Hartman et al. 2006) and exploited this unique opportunity of beat Cepheids and determined the metallicity gradient of M33 to be 0.16 dex kpc^{-1}, which is in good agreement with that derived from HII regions (Garnett et al. 1997). In the case of M31, we were able to detect 17 beat Cepheids with deep imaging and intra-night cadence from the PAndromeda survey (Lee et al. 2013). These 17 beat

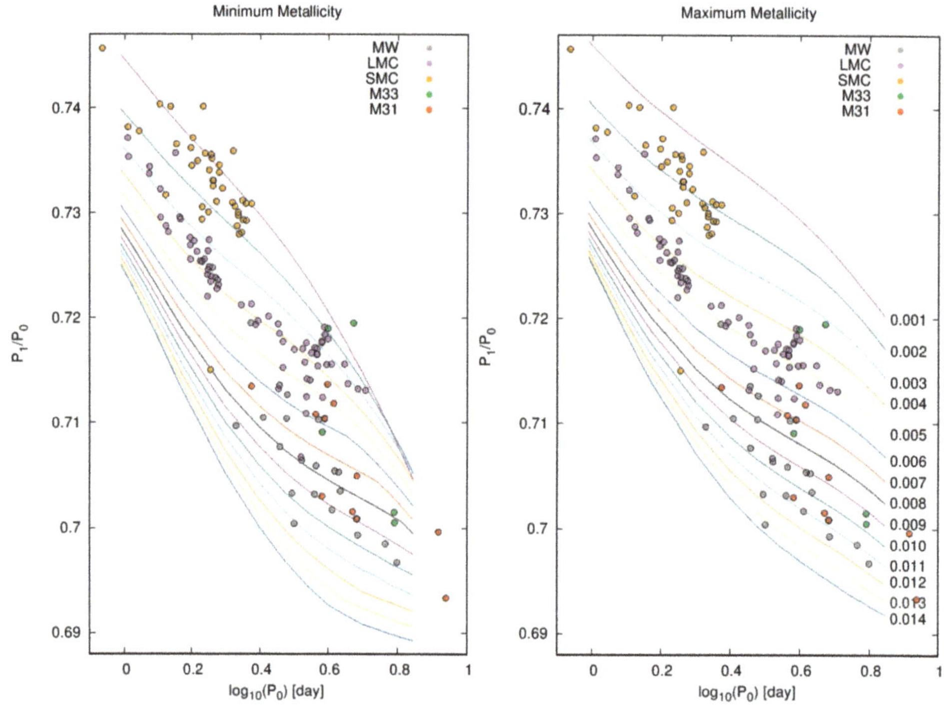

Figure 3.2. Metallicity estimates using beat Cepheids. From theoretical models, we can derive the minimum and maximum allowed metallicity given the period and period ratio. Beat Cepheids in the Magellanic Clouds (purple and orange circles) are relatively metal-poor compare to those in the Milky Way, M31 and M33 (gray, red, and green circles).

Cepheids also allow us to determine a metallicity gradient of 0.008 dex kpc^{-1}, in good agreement with that from planetary nebulae studies (Kwitter et al. 2012).

Another possibility of using Cepheids to determine metallicity is via modeling the bump in the light curves. It has been shown that when phase-folded the light curve of classical Cepheids, there will be a bump, aside from the light curve maximum, for the Cepheids with periods close to 10 days. This fine bump feature is caused by the resonances of the fundamental mode and the second overtone (see, e.g., Wood et al. 1997). In the Milky Way, the bump is more prominent in Cepheids with periods slightly longer than 10 days. On the other hand, in the Magellanic Clouds the Cepheids exhibit more significant bumps with periods slightly shorter than 10 days. This subtle difference in the bump Cepheids from different galaxies suggests that the bump is associated with the metallicity. In addition, this bump also caused the broken-law of the Leavitt Law that is close to 10 days. Keller & Wood (2006) have shown that by varying the metallicity in their modeled light curves, they can see the bump shifts to later phases with higher metallicity. This suggests that the bump Cepheids can be used to estimate the metallicity. However, the bump also depends on other model parameters, e.g., the effective temperature and the ratio of the fundamental mode to the second overtone, hence it has less constraining power than the beat Cepheids.

3.1.2 Eclipsing Binaries

Besides Cepheids, another commonly used class of periodic variable that can provide precise distance estimates is the eclipsing binary. Bohdan Paczynski (1997) first pointed out that eclipsing binaries can provide a single-step distance estimate to nearby galaxies to percent levels. The idea is that by combining both photometric and spectroscopic information, we can determine the fundamental properties of both stellar components of the binary system; when compare the visual magnitudes with that derived from synthetic stellar models, we can thus infer the distance. More specifically, the photometric measurements or light curves alone can provide a wealth of information on the physical configuration of the eclipsing binary system (see, e.g., Figure 3.3). This includes the inclination angle, the periods and the eccentricity of the orbit, the mass ratio of the two stellar components, and the individual size of the stellar components in the units of orbital distance. On the other hand, multi-epoch spectroscopic observations will provide radial velocity curves. The semi-amplitude of the radial velocity curve can provide the information on the absolute mass of each component and the orbital distance, hence deliver the masses and radii of each stellar component. In addition, from single-epoch spectra, we can determine the temperature of the stellar components when the absorption/emission

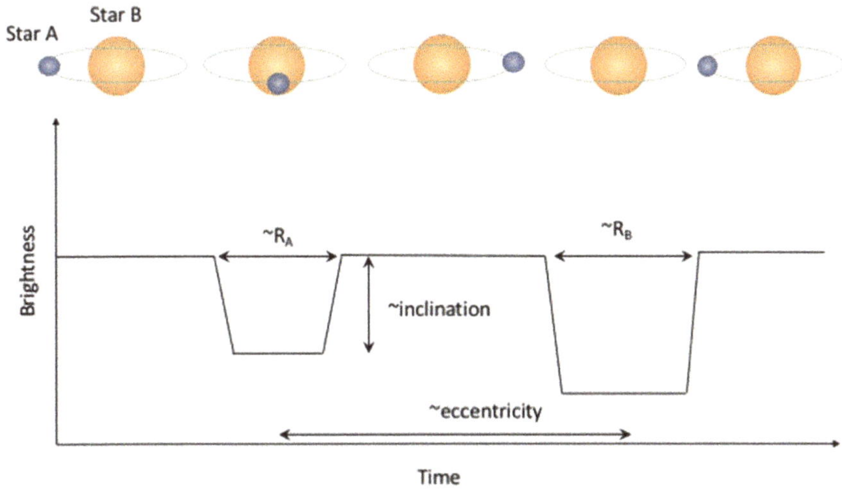

Figure 3.3. Depiction of an eclipsing binary system along with their brightness variation as viewed from an edge-on angle. We can estimate several parameters of the system configuration from the light curve alone. For example, the depth of the eclipses depends on the inclination angle of the orbit (the larger the inclination, the deeper the eclipse depth), but more important is the temperature contrast combined with the relative surface areas of the two components. The duration of the eclipses also depends on the orbital velocities and hence is correlated with the orbital period. The separation of the two eclipses provides information on the eccentricity. If they are equally separated by half of the orbital period, then the system is likely to have a circularized orbit. If the eclipses are unequally distributed across the orbital period, it is a sign of eccentric orbits.

features of each component are resolved. While there is a temperature–gravity degeneracy when modeling the spectra of a single star, it is no longer the case for eclipsing binaries. This is because with the radial velocity information we can determine the mass and radius independently, thus determine the temperature unambiguously. Even more so, when it comes to determine distances, there are several merits of eclipsing binaries compared to Cepheids. First, we can quantify the crowding/blending effects when fitting the light curves of eclipsing binaries, by introducing the third light ratio parameter. Please note that the crowding/blending effects are often hard to constrain for other distance indicators, especially at the extra-galactic distance where even the exquisite spatial resolution by space-based telescopes is not sufficient to resolve the distance indicators. Second, we can also determine the reddening law toward the eclipsing binaries using SED fitting (Bonanos et al. 2011). The reddening law is often hard to measure in the case of other distance indicators (this is why we prefer to use infrared Leavitt Law than that in the optical because the dust extinction effect is comparably negligible in the infrared). In this regard, we often assume a prior knowledge of the reddening law (e.g., $R_V = 3.1$ as similar in the Milky Way). But such reddening law may not be applicable to the local environment of the distance indicators. For example, Bonanos et al. (2011) investigated one eclipsing binary system in M33, and found an $R_V = 5.8$, which significantly deviates from the commonly assumed reddening law of $R_V = 3.1$. Once the concerns of crowding/blending and extinction have been removed, we can firmly infer the distance by matching the observed flux to that predicted from stellar modeling.

It is noted that conventional wisdom suggests that detached eclipsing binaries are more suitable for distance estimate. This is because they do not fill their Roche lobes, hence suffer less from deformation and are easier to predict their properties via synthetic stellar models. It has been shown that semi-detached eclipsing binaries can provide better constraining power than detached eclipsing binaries, especially as it is possible to unambiguously determine the mass ratio from the light curve (Wyithe & Wilson 2002). This is especially the case since we can firmly infer the mass ratio from light curve modeling of semi-detached binaries, which is otherwise insensitive to the mass ratio in the case of detached binaries (Wyithe & Wilson 2001, 2002).

It has been demonstrated that through semi-detached eclipsing binaries (see Figure 3.4), we can determine the M31 distance to a precision of 5% merely using one single eclipsing binary system (Ribas et al. 2005). The only disadvantage of this method is that at the distance of M31, even the brightest eclipsing binary systems ($V = 19 - 20$ mag) require more than 1 hr of integration time even with the state-of-the-art instruments on-board the 8–10 m telescope to reach sufficient signal-to-noise ratio to obtain a single epoch spectrum. This means we need hours of observations to acquire several epochs of spectroscopic observations to firmly determine the radial velocity curves. On top of that, there are not many bright eclipsing binaries known in M31 to-date. Indeed, Vilardell et al. (2006) only found 24 eclipsing binaries (including semi-detached) brighter than $V = 20.5$ mag from their deep INT observations. On the other hand, Lee et al. (2014b) only found 13 detached eclipsing binaries brighter than $V = 20.5$ mag from the PAndromeda survey. Please note that

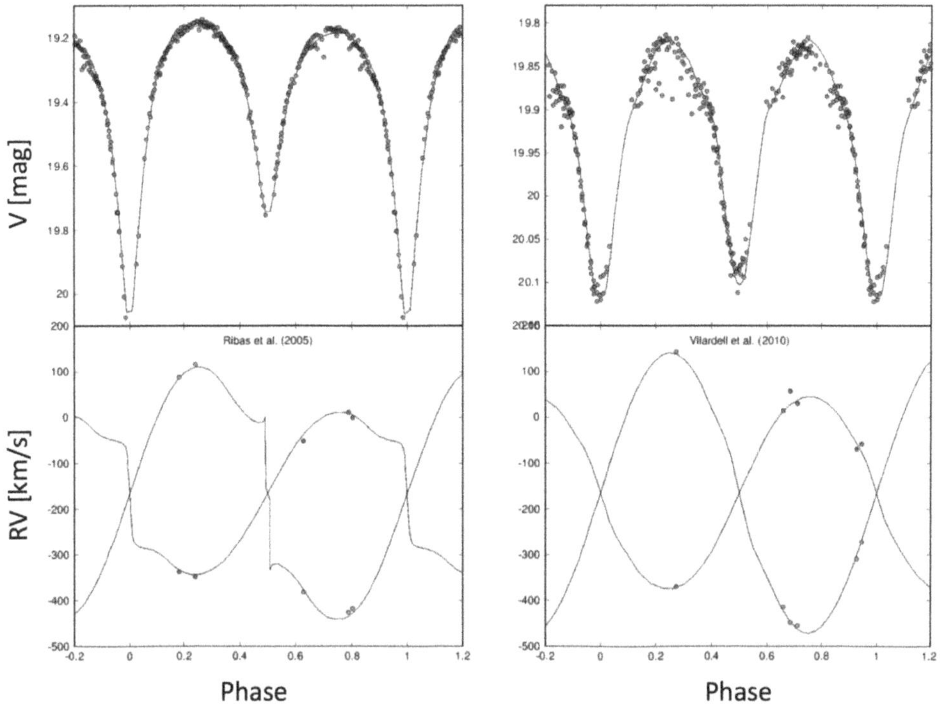

Figure 3.4. Light curve and radial velocity observations and best-fit models of two eclipsing binaries in M31. The left panel is from Ribas et al. (2005); the right panel is from Vilardell et al. (2010).

these eclipsing binaries are mostly composed of early-type (spectral type of O or A) stars, which require dedicated non-LTE modeling to extract physical parameters from their spectra. Other possibilities are the use of late-type stars that have a tight empirical color–size relation from interferometry, as shown by Pietrzynski et al. (2013). However, eclipsing binaries composed of late-type stars tend to have much longer periods, and they are also faint compared to those from early-type stars. In this regard, there have been hopes that by combining all the archival M31 data from different telescopes/surveys, we will be able to identify late-type eclipsing binaries' systems, providing complementary distance estimates from both early and late-type binaries.

3.2 Aperiodic Variables

3.2.1 Luminous Blue Variables

Massive stars (more massive than eight solar masses) are key players in galaxy evolution. They emit strong UV radiation, which heat the dust and fuel the far-infrared luminosity of galaxies (Massey 2013). They also exert a large amount of mechanical energy: through strong stellar winds and supernova explosion, they provide materials that enrich the galaxy chemically. Despite their importance, we do not have a full understanding of their evolution and populations at different stages due to their rarity. This is even more so for the elusive, short-lived phases of the luminous blue variables. During their long-term effort to observe M31 and M33 with the 100 inch telescope at Mount Wilson, Edwin Hubble and Allan Sandage (Hubble & Sandage 1953) found five irregular variables with high luminosities, blue colors that exhibit F-type spectroscopic features. Peter Conti (1984) was the first one to group these Hubble-Sandage variables with S Doradus variables, η Carina, and other similar variables. He coined the term luminous blue variables and separated this type of variable from Wolf-Rayet stars (see Figure 3.5).

The luminous blue variables are bright massive stars located at the upper left end of the H-R diagram. They undergo dramatic mass losses via episodic eruptions in

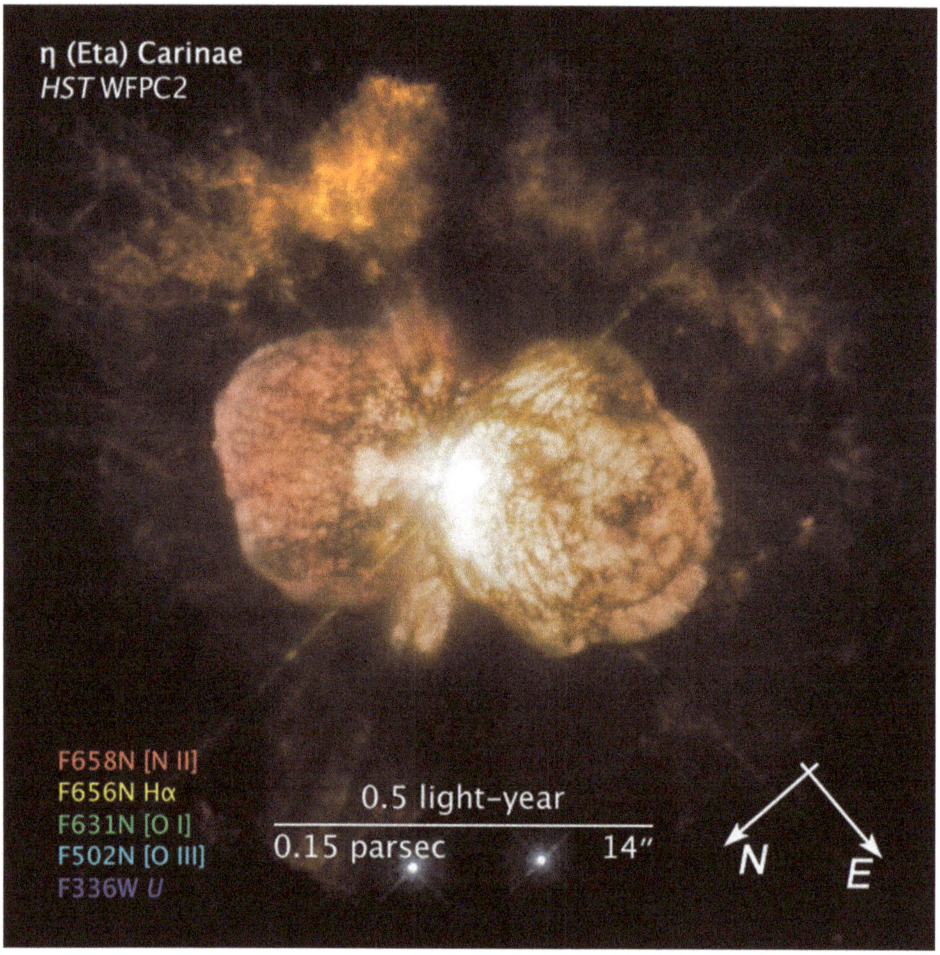

Figure 3.5. Composite image of η Carina from separate exposures made by the WFPC2 instrument on the *Hubble Space Telescope*. Five filters (F658N, F656N, F631N, F502N, F336W) were used to sample narrow wavelength ranges. The color results from assigning different hues (colors) to each monochromatic image. Please note the copious amount of materials around η Cariana, which can be associated with past eruptions. Image Credit: NASA, ESA, and the Hubble SM4 ERO Team.

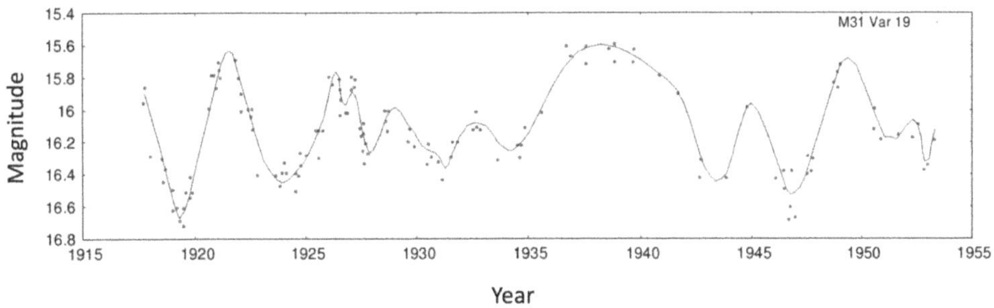

Figure 3.6. M31 Var 19, a luminous blue variables observed by Hubble & Sandage (1953).

the timescales of decades to a century (Humphreys & Davidson 1994). There are currently two contradicting scenarios to explain the luminous blue variables. From the perspective of stellar evolution modeling, they blow away very large portions of their mass during their eruptions, hence are regarded as a transitional phase when O stars evolve to hydrogen depleted Wolf-Rayet stars. The observational perspective, on the other hand, speculates that these luminous blue variables are the immediate precursor of type II supernovae, e.g., SN 1987A (Smith 2007), SN 2005gl (Gal-Yam & Leonard 2009), and SN 2009ip (Mauerhan et al. 2013). Due to their rarity (only 12 known Galactic luminous blue variables; Clark et al. 2005), the duration and properties of this short-lived luminous blue variable phase are still poorly understood. Although new ongoing infrared Galactic plane surveys are reporting several new candidates (Gvaramadze et al. 2010), it is difficult to have an accurate distance estimate, and follow them up individually is observationally very expensive. In this regard, one possible solution is to identify and characterize luminous blue variables in the local group galaxies instead. There are several advantages through this approach. First of all, compared to the individual luminous blue variables in the Milky Way, we have a better distance estimate to local group galaxies, and can use this to better infer the intrinsic luminosity of the luminous blue variables residing in those galaxies. Second, at the distance of local group galaxies, we can conveniently cover many luminous blue variables in the same galaxy with multi-object spectrographs, enabling efficient follow-up.

Because the luminous blue variables only exhibit large variabilities in the timescales of decades or a century (see Figure 3.6), it is difficult to discover them without a long-term photometric monitoring campaign. In this regard, there have been suggestions to identify luminous blue variables using other photometric and/or spectroscopic properties. For example, Massey et al. (2007) conducted a thorough survey of the entire M31 disk using both broad-band and narrow-band filters—H-alpha, SII, and OIII filters, in particular—to identify luminous blue variables via their strong H-alpha emissions and selected several candidates. They then followed up these candidates spectroscopically, and the spectra of these candidates exhibit similar features as bona fide luminous blue variables. The only missing evidence to establish the luminous blue variable nature is the lack of long-term light curves. On the other hand, Lee et al. (2014a) made use of the PAndromeda data and identified

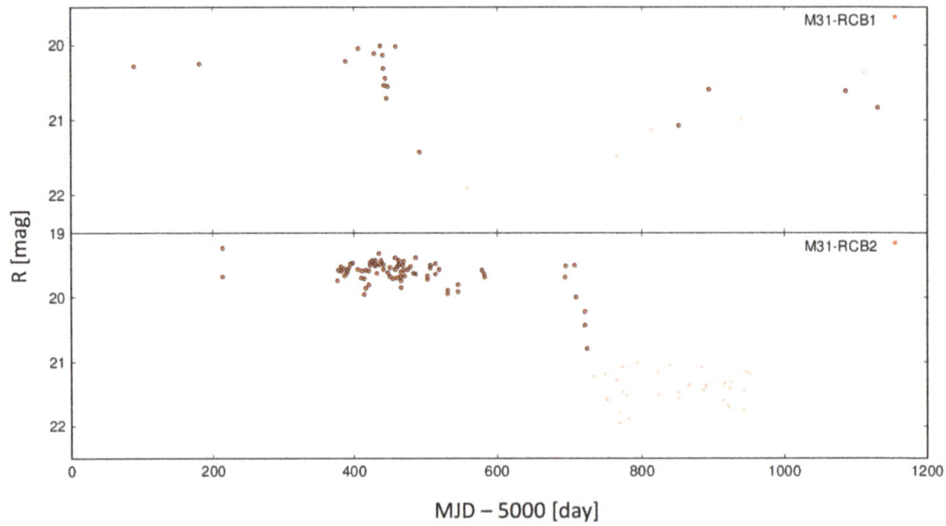

Figure 3.7. Light curves of two M31 R CrB stars from the Palomar Transient Factory (Tang et al. 2013). The red points are direct detections, while the pink points are upper limits set by non-detections.

possible candidates using short-term (and smaller) variations similar to known luminous blue variables. When we combine public archival M31 data from different telescopes, we can now form a baseline of two decades, enabling us to verify whether these candidates are true luminous blue variables via long-term variability.

3.2.2 R Coronae Borealis Stars

The R Coronae Borealis class (often dubbed RCB or R CrB) is composed of hydrogen deficient and carbon enriched supergiants with large aperiodic dimming, with durations spanning from weeks to years and dimming of up to nine magnitudes in their optical light curves (see Figure 3.7). Such large, aperiodic dimming is caused by sudden formation of carbon dusts that obscure the photosphere of the star. While R CrBs are very bright (absolute magnitude between −5 and −3.5), so far only 76 R CrBs are known in the Milky Way (Tisserand et al. 2011). This is due to our fixed position inside the Milky Way and the fact that there were no dedicated long-term all-sky survey sufficient to identify the sudden dimming of the R CrBs. Furthermore, the R CrBs may be in a short-lived, elusive phase of the stellar evolution, hence are difficult to identify.

Given the fact of hydrogen deficiency and carbon enrichment in the atmospheres of R CrBs, this suggest that R CrBs are at the end stage of stellar evolution. Currently there are two proposed mechanisms to explain the formation of R CrBs. The first scenario suggests that R CrBs are stars expanding rapidly (to the size of supergiants) right before turning into a white dwarf, as a final helium shell flash (so-called FF scenario; Iben et al. 1996). The second proposed scenario is that R CrBs are end products of white-dwarf merging events, where they formed by a carbon–oxygen white dwarf merging with a helium white dwarf (Webbink 1984). Since this scenario involves two degenerate stars (white-dwarfs), it is also referred to as the double degenerate (DD) scenario. From observations, there are several pieces of evidence supporting both these scenarios. First of all, observations found over-abundance of oxygen isotope ^{18}O and other abundance anomalies (e.g., fluorine) in the surface of cool R CrBs (Clayton et al. 2007), supporting the DD scenario. However, the discoveries of four R CrBs associated with nebulae (Clayton et al. 2011), in contrast, provided pieces of evidence that R CrBs may originate from the FF scenario. We note that R CrBs formed via the DD scenario are of special interests, because they can serve as low-mass analogs of SNe Ia formed via the DD channel. In this regard, studies of R CrBs might provide hints to the progenitor of SNe Ia.

Due to our fixed location inside the Milky Way and given the high extinction toward and through the Galactic bulge, it is difficult to carry out a complete census of R CrBs in the Milky Way. In this regard, M31—our closest neighboring spiral galaxy—provides an excellent opportunity to search for R CrBs with a complete view of an entire galaxy. This is especially the case because at the distance of M31, R CrBs are bright enough to be detected even with moderate size (1–2 m) telescopes. While we may not be able to pick up the signal of R CrBs during their dimming phases, we can simply identify R CrBs from their disappearance or re-appearance with long-term photometric surveys. Indeed, Tang et al. (2013) made use of the Palomar Transient Factory coverage of the M31 area to search for R CrBs, and found four candidates in M31. They further obtained spectroscopic follow-up and confirmed two of them to be bona fide R CrBs. With the public archival M31 data gathered from all over the globe, it is now possible to establish light curves of stars in the vast majority of the M31 disks (both north and south) with time baseline spanning two decades. This will enable us not only to search for R CrBs, but to conduct statistic studies and census of R CrBs in an external galaxy as well.

References

Alcock, C., Allsman, R. A., Axelrod, T. S., et al. 1995, AJ, 109, 1654

Beaulieu, J.-P., Buchler, J. R., Marquette, J.-B., Hartman, J. D., & Schwarzenberg-Czerny, A. 2006, ApJ, 653, L101

Bird, J. C., Stanek, K. Z., & Prieto, J. L. 2009, ApJ, 695, 874

Bonanos, A. Z., Castro, N., Macri, L. M., & Kudritzki, R.-P. 2011, ApJ, 729, L9

Bono, G., Marconi, M., Cassisi, S., et al. 2005, ApJ, 621, 966

Chen, X., Deng, L., de Grijs, R., Wang, S., & Feng, Y. 2018, ApJ, 859, 140

Clark, J. S., Larionov, V. M., & Arkharov, A. 2005, A&A, 435, 239

Clayton, G. C., Geballe, T. R., Herwig, F., Fryer, C., & Asplund, M. 2007, ApJ, 662, 1220

Clayton, G. C., Sugerman, B. E. K., Stanford, S. A., et al. 2011, ApJ, 743, 44

Conti, P. S. 1984, in IAU Symp. 105, Observational Tests of the Stellar Evolution Theory, ed. A. Maeder, & A. Renzini (Cambridge: Cambridge Univ. Press), 233

Davidge, T. J., McConnachie, A. W., Fardal, M. A., et al. 2012, ApJ, 751, 74

Eddington, A. S. 1917, Obs, 40, 290

Efstathiou, G. 2014, MNRAS, 440, 1138

Fiorentino, G., Clementini, G., Marconi, M., et al. 2012, Ap&SS, 341, 143

Freedman, W. L., & Madore, B. F. 2011, ApJ, 734, 46
Freedman, W. L., Madore, B. F., Gibson, B. K., et al. 2001, ApJ, 553, 47
Gal-Yam, A., & Leonard, D. C. 2009, Nature, 458, 865
Garnett, D. R., Shields, G. A., Skillman, E. D., Sagan, S. P., & Dufour, R. J. 1997, ApJ, 489, 63
Gordon, K. D., Bailin, J., Engelbracht, C. W., et al. 2006, ApJ, 638, L87
Gvaramadze, V. V., Kniazev, A. Y., & Fabrika, S. 2010, MNRAS, 405, 1047
Hartman, J. D., Bersier, D., Stanek, K. Z., et al. 2006, MNRAS, 371, 1405
Henden, A. A. 1979, MNRAS, 189, 149
Henden, A. A. 1980, MNRAS, 192, 621
Hubble, E., & Sandage, A. 1953, ApJ, 118, 353
Humphreys, E. M. L., Reid, M. J., Moran, J. M., Greenhill, L. J., & Argon, A. L. 2013, ApJ, 775, 13
Humphreys, R. M., & Davidson, K. 1994, PASP, 106, 1025
Iben, I. Jr., Tutukov, A. V., & Yungelson, L. R. 1996, ApJ, 456, 750
Ita, Y., Tanabé, T., Matsunaga, N., et al. 2004, MNRAS, 353, 705
Keller, S. C., & Wood, P. R. 2006, ApJ, 642, 834
Kodric, M., Riffeser, A., Hopp, U., et al. 2013, AJ, 145, 106
Kodric, M., Riffeser, A., Seitz, S., et al. 2015, ApJ, 799, 144
Kwitter, K. B., Lehman, E. M. M., Balick, B., & Henry, R. B. C. 2012, ApJ, 753, 12
Leavitt, H. S., & Pickering, E. C. 1912, HarCi, 173, 1
Lee, C.-H., Kodric, M., Seitz, S., et al. 2013, ApJ, 777, 35
Lee, C.-H., Seitz, S., Kodric, M., et al. 2014a, ApJ, 785, 11
Lee, C.-H., Koppenhoefer, J., Seitz, S., et al. 2014b, ApJ, 797, 22
Majaess, D., Turner, D., & Gieren, W. 2011, ApJ, 741, L36
Marquette, J. B., Beaulieu, J. P., Buchler, J. R., et al. 2009, A&A, 495, 249
Massey, P. 2013, NewAR, 57, 14
Massey, P., McNeill, R. T., Olsen, K. A. G., et al. 2007, AJ, 134, 2474
Mauerhan, J. C., Smith, N., Filippenko, A. V., et al. 2013, MNRAS, 430, 1801
Ngeow, C.-C., Lee, C.-H., Yang, M. T.-C., et al. 2015, AJ, 149, 66
Oosterhoff, P. T. 1957a, BAN, 13, 320
Oosterhoff, P. T. 1957b, BAN, 13, 317
Paczynski, B. 1997, in Proc. ST Scl Symp., The Extragalactic Distance Scale, ed. M. Livio, M. Donahue, & N. Panagia (Cambridge: Cambridge Univ. Press), 273
Perlmutter, S., Aldering, G., Goldhaber, G., et al. 1999, ApJ, 517, 565
Pietrzyński, G., Graczyk, D., Gieren, W., et al. 2013, Nature, 495, 76
Pike, C. D., & Andrews, P. J. 1979, MNRAS, 187, 261
Ribas, I., Jordi, C., Vilardell, F., et al. 2005, ApJ, 635, L37
Riess, A. G., Fliri, J., & Valls-Gabaud, D. 2012, ApJ, 745, 156
Riess, A. G., Filippenko, A. V., Challis, P., et al. 1998, AJ, 116, 1009
Riess, A. G., Macri, L. M., Hoffmann, S. L., et al. 2016, ApJ, 826, 56
Riess, A. G., Casertano, S., Yuan, W., et al. 2018, arXiv:1801.01120
Smith, N. 2007, AJ, 133, 1034
Soszynski, I., Udalski, A., Szymanski, M., et al. 2000, AcA, 50, 451
Tang, S., Cao, Y., Bildsten, L., et al. 2013, ApJ, 767, L23
Tisserand, P., Wyrzykowski, L., Wood, P. R., et al. 2011, A&A, 529, A118
Udalski, A., Soszynski, I., Szymanski, M., et al. 1999, AcA, 49, 1

Vilardell, F., Jordi, C., & Ribas, I. 2007, A&A, 473, 847
Vilardell, F., Ribas, I., & Jordi, C. 2006, A&A, 459, 321
Vilardell, F., Ribas, I., Jordi, C., Fitzpatrick, E. L., & Guinan, E. F. 2010, A&A, 509, A70
Webbink, R. F. 1984, ApJ, 277, 355
Wood, P. R., Arnold, A. S., & Sebo, K. M. 1997, ApJ, 485, L25
Wyithe, J. S. B., & Wilson, R. E. 2001, ApJ, 559, 260
Wyithe, J. S. B., & Wilson, R. E. 2002, ApJ, 571, 293
Zhevakin, S. A. 1963, ARA&A, 1, 367

Chien-Hsiu Lee

Chapter 4

Transients

Toward the lines-of-sights of M31, there can be different types of transients. They can be intrinsic bright phenomena, such as microlensing events and/or nova outbursts that appear to be faint at the distance of M31. They can also be intrinsic faint phenomena, e.g., dwarf novae, stellar flares, and FU Ori/EXor variables, which are actually lying in the foreground and are associated with the Milky Way rather than M31. Here, we focus on the transient phenomena that are at the distance of M31, and how these phenomena can help us understand the properties (or the underlying stellar populations) of M31.

4.1 Microlensing

4.1.1 Brief History

Back in the 1930s, when estimating the mass of the Coma galaxy cluster using the Virial theorem, Fritz Zwicky (1933) noticed that there are large fractions of unseen matter, which he first proposed the idea and coined the term "Dark matter." The existence of dark matter is further supported by spectroscopic observations of M31 by Vera Rubin and Kent Ford Jr., where they found that the rotation curve did not drop even at the outskirts of M31, hinting to the existence of a large amount of unseen matter (Rubin & Ford 1970). However, eight decades after the discovery work by Fritz Zwicky, the nature of dark matter is still unknown. There are several candidates proposed for dark matter, though. It can be in the form of massive compact objects in the halo (MACHOs), or it can also be smoothly distributed, i.e., weakly interacting massive particles (WIMPs). Because dark matter hardly exposes its traces with electro-magnetic emissions, an elegant method to search for and characterize dark matter is via gravitational interaction. Albert Einstein (1936) was the first to derive the gravitational lensing formula of compact objects (MACHOs) composed of stellar masses and suggested that observing such events are unlikely. Nevertheless, Bohdan Paczynski (1986) further conducted detailed calculations and

doi:10.1088/2514-3433/ab3639ch4

concluded that while the chance (optical depth) is relatively low—at one in a million when a background is lensed and amplified by a factor of 1.34 by MACHOs in the Milky Way halo—it is still possible to use gravitational lensing to search for MACHOs if we can observe more than a million stars at a time. As the lens masses (of MACHOs) are in the range of stellar masses, the images of the background source will be separated in the order of micro-arcseconds, hence the name microlensing. The beauty of microlensing is that the light curve shape is symmetrical, hence can be easily distinguished from other transients and/or variable stars.

4.1.2 Theory

Microlensing formalism can be easily derived using Figure 4.1. Using the triangle formed by the observer (O), background source (S), and the lensed image (I), the length between the image and the source can be expressed as

$$\bar{IS} = \alpha D_{LS} = (\theta_I - \theta_S) D_{OS}, \tag{4.1}$$

where $D_{OS} = D_{OL} + D_{LS}$ is the distance between the observer and the background source. The light bending angle α can be calculated from the general theory of relativity as

$$\alpha = \frac{4GM}{c^2 D_{OL} \theta_I}, \tag{4.2}$$

where G and c are the gravitational constant and the speed of light, respectively. M denotes the underlying mass of the lens. Using this equation, we can then derive the lens equation:

$$\theta_I(\theta_I - \theta_S) = \frac{4GM}{c^2} \frac{D_{LS}}{D_{OL} D_{OS}} := \theta_E^2, \tag{4.3}$$

where θ_E denotes the angular Einstein ring radius. When the observer, the lens, and the background source are well-aligned on a straight line ($\theta_S = 0$), the image from the background source will become a ring shape, which is dubbed the Einstein ring. The angular Einstein ring radius is

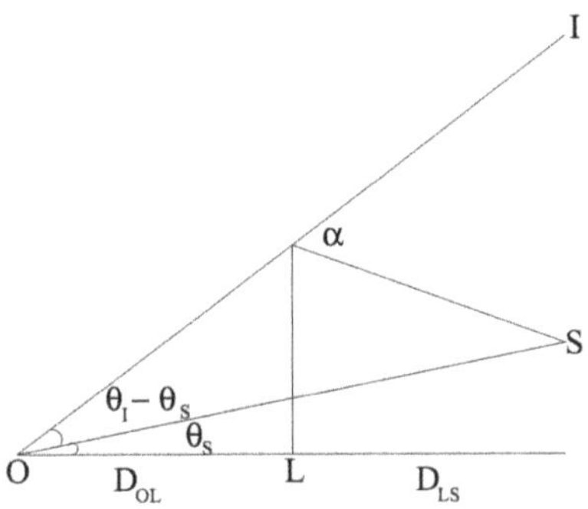

Figure 4.1. Gravitational lensing formalism.

$$\theta_E = \sqrt{\frac{4GM}{c^2}\left(\frac{1}{D_{OL}} - \frac{1}{D_{OS}}\right)} \tag{4.4}$$

$$= 0.902\ mas \left(\frac{M}{M_\odot}\right)^{1/2}\left(\frac{10kpc}{D_{OL}}\right)^{1/2}\left(1 - \frac{D_{OL}}{D_{OS}}\right)^{1/2}. \tag{4.5}$$

If we express all length scales in terms of the θ_E, and define the source angular position as

$$u := \theta_S/\theta_E, \tag{4.6}$$

then by solving the lens equation, we can determine the position of the two lensed images:

$$u_\pm = (u \pm \sqrt{u^2 + 4})/2, \tag{4.7}$$

where $u_\pm = \theta_{I_\pm}/\theta_E$. Because of the conservation of surface brightness, the magnification factor of the background source can be easily computed by comparing the area of the lensed images to the area of the unlensed source. So, the magnification factor for each of the lensed images, as well as the total magnification, can be derived as

$$A_\pm = \left|\frac{u_\pm}{u}\frac{du_\pm}{du}\right|, \tag{4.8}$$

$$A_{\text{total}} = A_- + A_+ \tag{4.9}$$

$$= \frac{u^2 + 2}{u\sqrt{u^2 + 4}}. \tag{4.10}$$

When u $\ll$ 1, A_{total} can be approximated by $1/u$. This is the beauty of microlensing because we can easily derive the light curve using the relative position between the lens and the source projected on the sky. For most of the cases, we can assume the relative motion between the lens and the source to be rectilinear, which allows us to decompose u into parallel and perpendicular components of the relative lens–source motion as a function of time:

$$u(t) = \sqrt{\left(\frac{t - t_0}{t_E}\right)^2 + u_0^2}, \tag{4.11}$$

and

$$A(t) = \frac{u(t)^2 + 2}{u(t)\sqrt{u(t)^2 + 4}}, \tag{4.12}$$

where t_0 and u_0 are the timing and the impact parameter when the lens is closest to the source along our line-of-sight. Here we introduce a new term, t_E, which is the

Einstein timescale that is defined by the time required for the lens to traverse one Einstein radius:

$$t_E = 0.214 \text{ year}\left(\frac{M}{M_\odot}\right)^{1/2}\left(\frac{D_{OL}}{10\text{kpc}}\right)^{1/2}\left(1 - \frac{D_{OL}}{D_{OS}}\right)^{1/2}\left(\frac{200\text{kms}^{-1}}{V_{\text{rel}}}\right). \tag{4.13}$$

With the advent of CCDs, and encouraged by Paczynski's seminal paper, several projects started to patrol dense stellar fields such as the Galactic bulge and both Magellanic Clouds with wide-field cameras in the 1990s, e.g., MACHO, EROS, OGLE, and MOA teams, to hunt for microlensing events and use them to study dark matter in compact form. Following these intense monitoring programs, the first microlensing events have been reported by Alcock et al. (1993), Aubourg et al. (1993), and Udalski et al. (1993). The results of these projects, however, are inconsistent. For example, after 5.7 years of continuous monitoring, the MACHO team reported 13 microlensing events toward the Large Magellanic Cloud (Alcock et al. 2000). While later analysis by David Bennett (2005) showed that three of them are (candidate) variables. Bennett (2005) then concluded that with the remaining 10 microlensing events, only up to 16% of the Milky Way halo is composed of MACHOs with masses between 0.1 and 1 solar masses. The EROS results, on the other hand, suggested that less than 8% of the Milky Way halo is composed of MACHOs with masses of 0.4 solar masses; they also ruled out MACHOs with masses between 0.6×10^{-7} and 15 solar masses as the major component of the Milky Way halo (Tisserand et al. 2007). In addition, results from the OGLE team (Wyrzykowski et al. 2010, 2011a, 2011b) indicated that microlensing events identified in both Magellanic Clouds are consistent with self-lensing, meaning that there is no need for MACHOs in the Milky Way halo to explain the microlensing events. However, recent re-analysis of the OGLE and EROS data by Novati et al. (2013) suggested that, on the contrary, some of the OGLE events can be attributed to lenses in the Milky Way halo.

4.1.3 Microlensing Observations in M31

One possible cause of the contradicting and inconclusive results among the MACHO, EROS, and OGLE teams is that given our fixed position inside the Milky Way, we are limited to a single line-of-sight and only probe a small fraction of the Milky Way halo. In addition, at the proximity of Magellanic Clouds, their three-dimensional structures are negligible, further complicating the interpretation of the observed microlensing events. In this regard, M31—our closest neighboring spiral galaxy—might be an ideal target for microlensing searches. This is because we can have multiple sight-lines toward the M31 bulge and disk through its dark halo, as suggested by Crotts (1992), Baillon et al. (1993), and Jetzer (1994). In addition, the properties of M31 are well-known. At the distance of 752 kpc, the internal structure of M31 is negligible and we can assume all of the sources are at the same distance. With a high inclination angle (inclination = 77°; Walterbos & Kennicutt 1987), we will see asymmetry of the microlensing event rates between the near side (north west) and the far side (south east) of its disk. While such effects can also originate from varying dust extinction at different sight-lines, it can be properly accounted for with a dedicated extinction map across the entire disk of M31 (Montalto et al. 2009). It is important to note that microlensing light curves in M31 differ significantly from that in the Milky Way and in the Magellanic Clouds. This is because at the great distance of M31, we cannot resolve the sources of the microlensing events, as each pixel of the detector can contain hundreds of stars. On the one hand, we can only extract pixel-based light curves, and the baseline flux of the source is unknown. On the other hand, since the microlensing signal stands out from the heavily blended source lights in one pixel, we can assume that it is a high magnification event, i.e., very small impact parameter where $u \ll 1$. In this case, we can simplify the light curve using the high magnification approximation formula by Gould (1996), which facilitates our searches in the big amount of pixellated light curves. The light curve or flux excess of the microlensing event can then be expressed as

$$\Delta_F \sim F_{\mathrm{eff}}\left(\frac{(t - t_0)^2}{t_{\mathrm{eff}}^2} + 1\right)^{-1/2}, \tag{4.14}$$

where in the pixel lensing regime, we can only detect the flux excess (Δ_F), and fit the light curves with the effective flux $F_{\mathrm{eff}} := \frac{F_0}{u_0}$ as a function of the intrinsic source flux F_0, and the effective timescale $t_{\mathrm{eff}} := u_0 t_E$.

Searches of microlensing events in M31 have started as early as the 1990s, where Tomaney & Crotts (1996) first used the Vatican Advanced Technology Telescope and the 4 m Mayall telescope at the Kitt Peak National Observatory to conduct the Columbia microlensing survey and reported six candidates, though some of them turned out to be variables with further continuous observations. Ansari et al. (1999) formed the AGAPE project and used the 2 m Bernard Lyot telescope at the French Pyrenees to discover one bright and short microlensing event. The AGAPE project was succeeded by the POINT-AGAPE project, which employed the 2.5 m Isaac Newton Telescope with its half-degree WFCam to monitor M31 from 1999 till 2002. The POINT-AGAPE project resulted in a handful of events reported independently by different working groups (Belokurov et al. 2005; Calchi Novati et al. 2005; Tsapras et al. 2010). Using the very same INT/WFCam data, the MEGA collaboration (de Jong et al. 2006) found 14 events. However, the MEGA collaboration concluded that these events are in good agreement with self-lensing events or variables and there is no need for MACHOs in the halo of M31 to explain the microlensing events.

The contradicting results may originate from the small sample of microlensing events in M31. To tackle the issue of a small sample, we will need either very long-term monitoring of M31 or a very wide-field of view with larger aperture to identify numerous microlensing events. Regarding the first prospect, Riffeser et al. (2001) started a decade-long monitoring of the M31 bulge for microlensing events from 1997 to 2008. Though only focused on the central $16' \times 16''$ region of M31 with a rather small (0.8 m) telescope, their 11 yr monitoring campaign yielded 12 microlensing events. Among these 12 events, two of them are highly magnified (see Figure 4.2) and are hard to be accounted for by self-lensing, thus requires MACHOs in the halo of M31 (Lee et al. 2015). As for monitoring with wider field-of-views and larger apertures, the PAndromeda project (Lee et al. 2012a) has discovered six events in the inner 20 × 20 arcminutes region of M31 using only 1 year of data. This is twice or thrice more events compared with other M31 microlensing projects, thanks to the very high cadence and larger aperture of the Pan-STARRS telescope. The full PAndromeda data set is still under analysis, and hopefully will provide us with a large enough sample to distinguish self-lensing from MACHO lensing, and deliver an upper limit of the fraction of MACHOs in the halo of M31.

Besides searching for dark matter candidates, microlensing is also a powerful tool to detect exoplanets (Mao & Paczynski 1991). Indeed, with the inconclusive results of MACHOs in the Milky Way halo, the microlensing campaigns, e.g., OGLE and MAO, have turned their attention to exoplanets. The merit of the microlensing method, compared to other exoplanet searches, is that the planet signal is neither associated with the planetary electromagnetic emission nor to that from the host star. In this regard, the microlensing method can probe exoplanets that are well beyond the snow line, with masses as small as Earth or Jovian satellites, that are otherwise not reachable with other exoplanet detection methods (see, e.g., Gaudi 2012 for a review). This is also true for exoplanets outside of our Milky Way and even those residing in M31. By re-analyzing the binary microlensing event POINT-AGAPE N2, Ingrosso et al. (2009) suggested that this microlensing event can be interpreted as a putative exoplanet system (see Figure 4.3), with a mass ratio of 0.014 to the primary lens (the host star), or a 6× Jovian mass exoplanet assuming

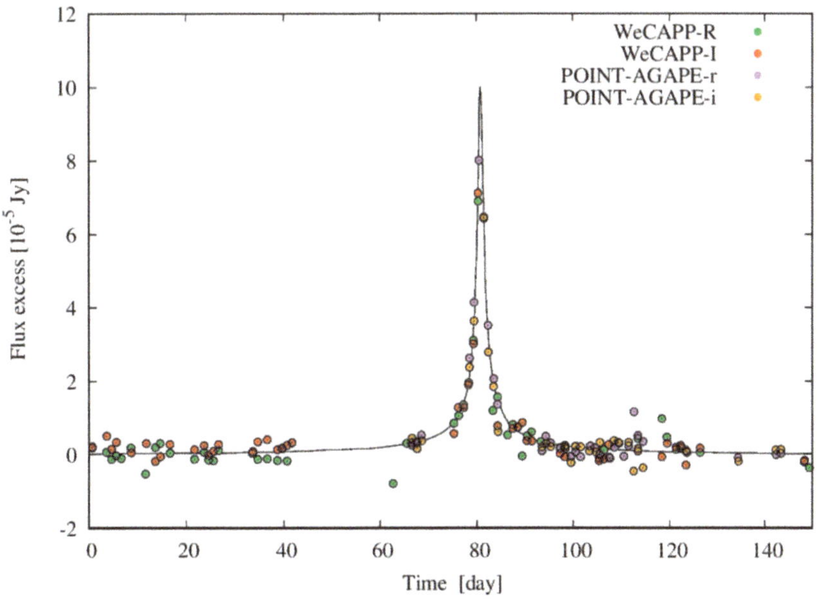

Figure 4.2. Light curve of M31 microlensing event WeCAPP-1/POINT-AGAPE-S3. Such high magnification events are in favor of MACHO lensing rather than self-lensing (Riffeser et al. 2008).

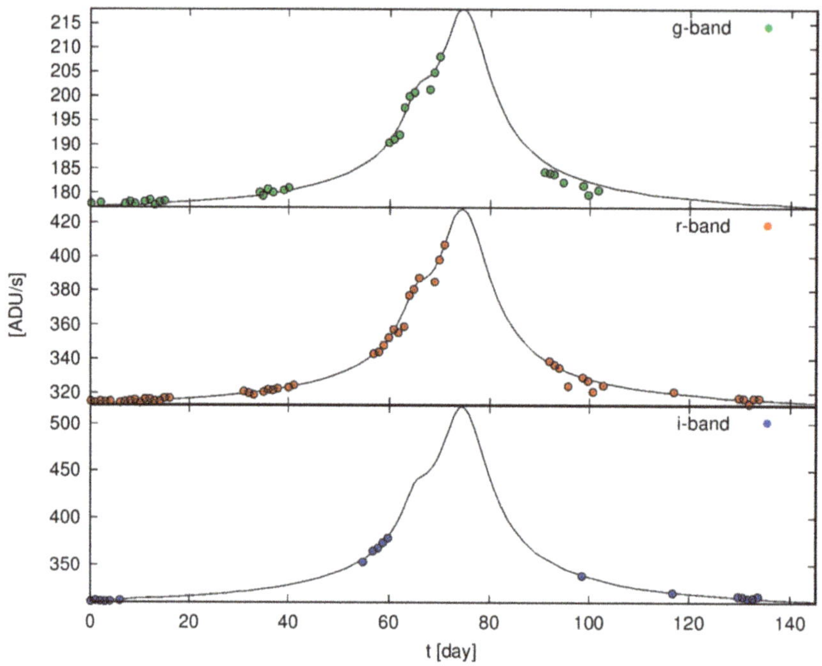

Figure 4.3. The POINT-AGAPE-99-N2 microlensing event, a binary lens event with the secondary lens to be a candidate planetary mass object.

the primary lens has solar properties. However, there are other possibilities, including binary stars or parallax effects, that also fit comparatively well to the light curves. Since POINT-AGAPE, there have also been other attempts to search for planetary microlensing in M31, e.g., H Dong et al. but there has not yet been a clear detection. One of the possible causes of the non-detection is that exoplanetary microlensing signature is less prominent than the stellar/MACHO microlensing signature. Take the exoplanetary microlensing in the Galactic bulge as an example, the stellar microlensing feature usually sustains for months, while the planetary signature only last for a couple of days. In addition, the magnification is proportional to the underlying mass, hence the planetary signature will be much weaker than the stellar microlensing magnification. However, there are two different types of planetary microlensing depending on the configuration of the caustic crossing. The first one is due to the planetary caustic, which leads to a planetary microlensing signature that is isolated and further away from the main stellar microlensing feature in time and requires very high cadence to discover. The second one is induced by central caustic crossing, in which the planetary signature is embedded during the peak of the stellar microlensing magnification, and is relatively easier to identify thanks to the highly magnified stellar microlensing peak in the light curves. As the time duration of the exoplanet microlensing signature in M31 is very short (less than a day), it will require dedicated, high-cadence monitoring with medium to large aperture telescopes with wide field-of-view, e.g., CFHT/MegaCAM or Subaru/HSC to achieve the required S/N and cadence in the future.

4.2 Novae

Novae span a subclass of cataclysmic variables that are composed of a white dwarf (as the primary star) accreting materials from a main-sequence or evolved star (as the secondary star, see Figure 4.4). The secondary star is believed to be non-degenerate, either on the main-sequence or at the late stage of stellar evolution; it is expected to expand and fill the Roche lobe, where the material are transferred via Roche-lobe overflow to the primary white dwarf. The white dwarf keeps accreting material until it reaches thermo-nuclear runaway at the surface and leads to the nova eruption. During eruption, the novae can be as bright as 15 magnitude visually in M31 and can be easily identified even with small aperture by amateur astronomers.

Novae are important in several aspects. First of all, Edwin Hubble first noticed that brighter novae decline faster (Hubble 1929). This led to the empirical relation of the Maximum Magnitude versus Rate of Decline (the so-called MMRD-relation), as shown in McLaughlin (1945) and Arp (1956), which can be used as a distance indicator. The theoretical foundation of MMRD-relation was established by Shara (1981) and Livio (1992).

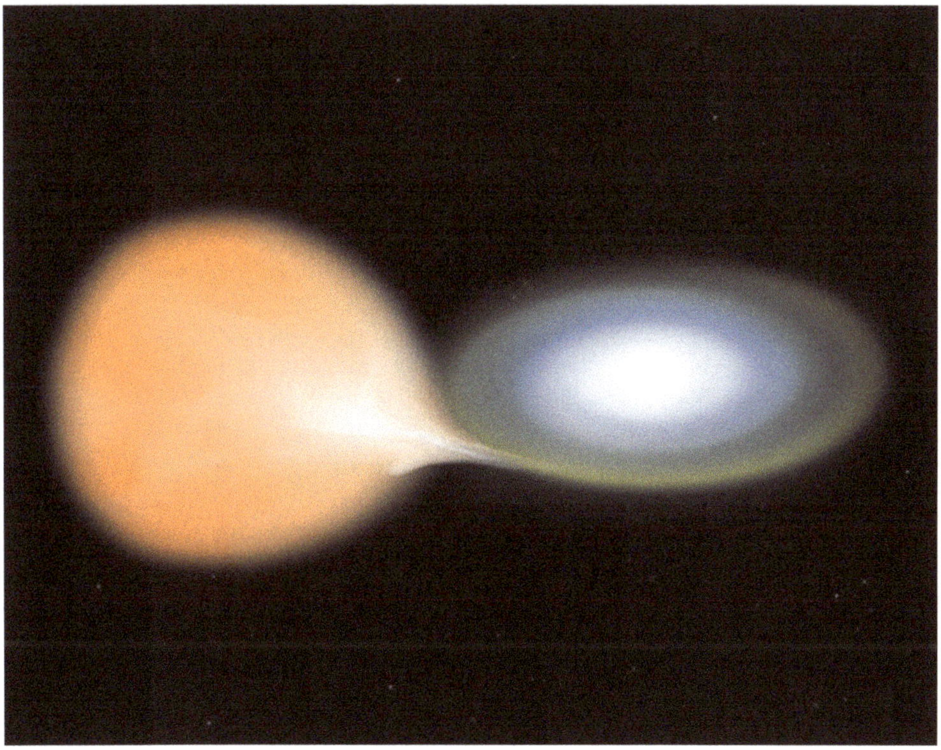

Figure 4.4. Artist's illustration depicting a classical nova binary system before an explosion on the surface of the white dwarf. Credit: (Illustration) NASA/CXC/M. Weiss.

Second, novae can serve as stellar population tracers. This is because as the central white dwarf becomes more massive, the decline of the nova light curve becomes steeper. To quantify the rate of decline, the term t_2 is commonly used to define the amount of time that the nova brightness declines by 2 magnitudes after the light curve maximum (see, e.g., Gaposchkin 1957). On top of that, della Valle & Livio (1995) further used t_2 to classify novae into fast ($t_2 < 12$ days) and slow ($t_2 > 12$ days) classes. After examining the spectra of different classes of novae, they found that fast novae are associated with Population I stars, i.e., the central white dwarf is more massive, while the slow novae are related to Population II stars, i.e., the central white dwarf is less massive.

Last but not least, novae are candidate progenitors of supernovae (Schaefer 2010). This is especially for a subclass of novae called recurrent novae, particularly when the central WD has a CO composition. It has been shown that recurrent novae can undergo several nova eruption, with eruption intervals as short as a year, like the case of M31N-2008-12a (Darnley et al. 2014, Tang et al. 2014), although it is not clear if the central white dwarf of M31N-2008-12a is a CO white dwarf. It has been proposed that recurrent novae might not burn out all the accreted materials after each eruption, and thus gain weight after several eruptions. This means that the central white dwarf will accumulate enough mass and eventually turn into a supernova explosion. Because recurrent novae at extra-galactic distance are difficult to identify, and we need to wait for years to decades to establish the recurrent nature, there has not been a smoking gun of a nova turned into a supernova yet.

Due to our fixed position in the Milky Way, it is difficult to provide a complete chart of novae in the Milky Way. This is especially the case for the part of the Milky Way that is heavily obscured behind the Galactic bulge. To make it worse, novae are heavily obscured by the dust produced during the eruption, hence it is difficult to determine their distance. Therefore, it has been proposed to search for novae in extra-galactic fields, especially in M31. On the one hand, M31 is close enough such that even a small aperture telescope can discover novae in M31. Indeed, previous discoveries of M31 novae before dedicated all-sky surveys are mostly conducted by amateur astronomers. Another advantage is that at the proximity of M31, we can resolve individual stars with the aid of the exquisite spatial resolution of the *Hubble Space Telescope*, hence conduct a detailed study of the progenitor of novae. On the other hand, M31 is also far enough that we can have a complete picture of its disk, enabling us to probe novae in different parts of M31, including its bulge, disk, and even the halo.

As mentioned before, most of the M31 novae were charted by the great contribution from amateur astronomers before dedicated all-sky surveys. There were many focused surveys to find novae in M31, e.g., Arp (1956), Rosino (1964, 1973, 1989), Ciardullo et al. (1987), Shafter & Irby (2001). More recent samples of M31 novae also came from the microlensing surveys in M31, where the high-cadence monitoring is not only useful for microlensing, but essential for other transient phenomena as well. The POINT-AGAPE survey, for example, has reported 20 novae in the disk of M31 (Darnley et al. 2004). The WeCAPP survey, concentrating on the central part of M31, also reported 91 novae in the M31 bulge (Lee et al. 2012b).

A working group at the Max Planck Institute for Extraterrestrial (Pietsch et al. 2007) has compiled a detailed list of M31 novae, including the location, date of eruption, speed and spectroscopic class, as well as light curves and spectra (if available).[1]

The microlensing data are pivotal for novae studies. For example, Lee et al. (2012b) used the WeCAPP data to study the light curves of 91 novae in M31. With the exquisite cadence of WeCAPP, they were able to not only derive the rate of decline, but classify the light curves using the morphology advocated by Strope et al. (2010). This includes smooth decline light curves (S-type), light curves with one single cusp due to a large amount of dust (C-type), light curves with jitters that are associated to the instabilities in the hydrogen burning envelopes (J-class), and light curves with features of series of quasi-sinusoidal oscillations that are linked to the strong magnetic field at the surface of a white dwarf (O-class). A gallery of these different type of light curves can be seen in Figures 4.5–4.7.

With the completion of the Panchromatic Hubble Andromeda Treasury (Dalcanton et al. 2012), the study of the progenitors of M31 novae is within reach. For example, Williams et al. (2014) have conducted a pilot study of a handful of novae in M31 combining high spatial resolution progenitor imaging from *HST*, and follow-up nova light curves and spectra on the ground. They concluded that M31 nova progenitors are composed excessively with red giant stars as secondaries, which are not seen in the Milky Way. Ongoing and future surveys of M31, e.g., the Zwicky Transient Factory (ZTF), will deliver a larger sample of novae in M31. By combining M31 novae from ZTF with the pre-eruption imaging from the *HST* data, we will have a more clear picture on the progenitors of novae.

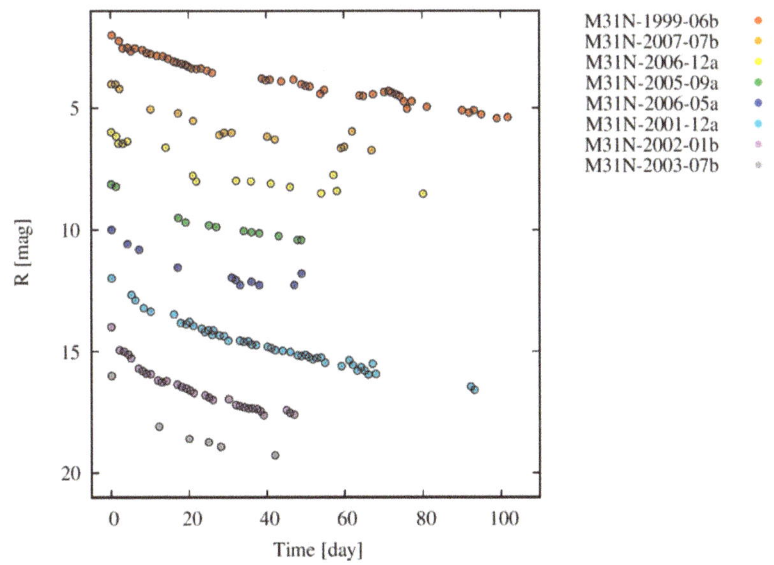

Figure 4.5. Light curves of smooth class novae in M31. Data from Lee et al. (2012b).

[1] For details, please see http://www.mpe.mpg.de/m̃31novae/opt/m31/index.php.

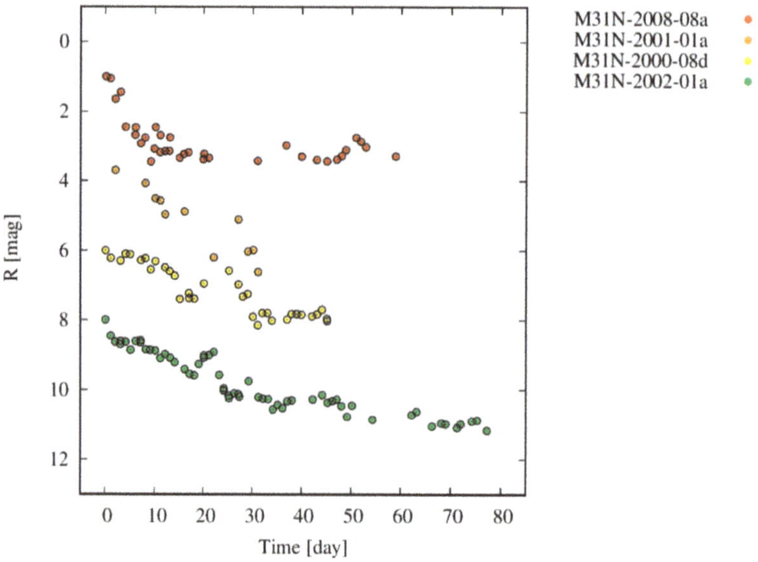

Figure 4.6. Light curve of M31 novae showing cusp features. Data from Lee et al. (2012b).

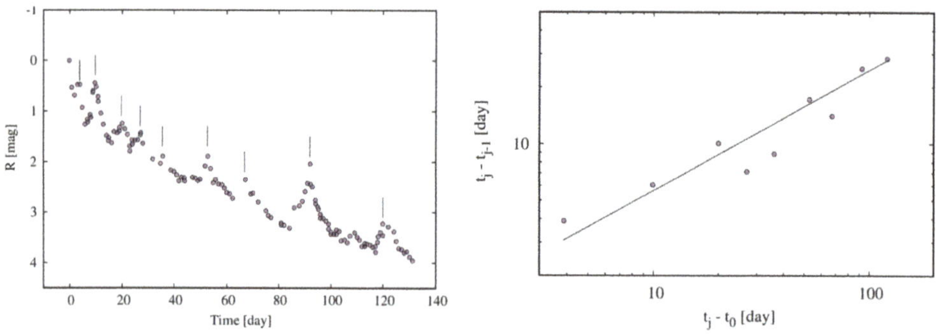

Figure 4.7. Upper panel: Light curve of a M31 nova showing jitter features. Lower panel: Time-interval between successive jitters versus time-interval of each jitter from the nova eruption. The trend among these two time-intervals is associated with the hydrogen burning envelope instabilities mechanism.

References

Alcock, C, Akerlof, C. W., Allsman, R. A., et al. 1993, Nature, 365, 621
Alcock, C., Allsman, R. A., Alves, D. R., et al. 2000, ApJ, 542, 281
Ansari, R., Aurière, M., Baillon, P., et al. 1999, A&A, 344, L49
Arp, H. C. 1956, AJ, 61, 15
Aubourg, E., Bareyre, P., Bréhin, S., et al. 1993, Nature, 365, 623
Baillon, P., Bouquet, A., Giraud-Heraud, Y., & Kaplan, J. 1993, A&A, 277, 1
Belokurov, V., An, J., Evans, N. W., et al. 2005, MNRAS, 357, 17
Bennett, D. P. 2005, ApJ, 633, 906
Calchi Novati, S., Mirzoyan, S., Jetzer, P., & Scarpetta, G. 2013, MNRAS, 435, 1582
Calchi Novati, S., Paulin-Henriksson, S., An, J., et al. 2005, A&A, 443, 911

Ciardullo, R., Ford, H. C., Neill, J. D., Jacoby, G. H., & Shafter, A. W. 1987, ApJ, 318, 520
Crotts, A. P. S. 1992, ApJ, 399, L43
Dalcanton, J. J., Williams, B. F., Lang, D., et al. 2012, ApJS, 200, 18
Darnley, M. J., Bode, M. F., Kerins, E., et al. 2004, MNRAS, 353, 571
Darnley, M. J., Williams, S. C., Bode, M. F., et al. 2014, A&A, 563, L9
de Jong, J. T. A., Widrow, L. M., Cseresnjes, P., et al. 2006, A&A, 446, 855
della Valle, M., & Livio, M. 1995, ApJ, 452, 704
Einstein, A. 1936, Science, 84, 506
Gaposchkin, C. H. P. 1957, The Galactic Novae (Amsterdam: North-Holland)
Gaudi, B. S. 2012, ARA&A, 50, 411
Gould, A. 1996, ApJ, 470, 201
Hubble, E. P. 1929, ApJ, 69, 103
Ingrosso, G., Novati, S. C., de Paolis, F., et al. 2009, MNRAS, 399, 219
Jetzer, P. 1994, A&A, 286, 426
Lee, C.-H., Riffeser, A., Seitz, S., Bender, R., & Koppenhoefer, J. 2015, ApJ, 806, 161
Lee, C.-H., Riffeser, A., Koppenhoefer, J., et al. 2012a, AJ, 143, 89
Lee, C.-H., Riffeser, A., Seitz, S., et al. 2012b, A&A, 537, A43
Livio, M. 1992, ApJ, 393, 516
Mao, S., & Paczynski, B. 1991, ApJ, 374, L37
Mclaughlin, D. B. 1945, PASP, 57, 69
Montalto, M., Seitz, S., Riffeser, A., et al. 2009, A&A, 507, 283
Paczynski, B. 1986, ApJ, 304, 1
Pietsch, W., Haberl, F., Sala, G., et al. 2007, A&A, 465, 375
Riffeser, A., Seitz, S., & Bender, R. 2008, ApJ, 684, 1093
Riffeser, A., Fliri, J., Gössl, C. A., et al. 2001, A&A, 379, 362
Rosino, L. 1964, AnAp, 27, 498
Rosino, L. 1973, A&AS, 9, 347
Rosino, L., Capaccioli, M., D'Onofrio, M., & della Valle, M. 1989, AJ, 97, 83
Rubin, V. C., & Ford, W. K. Jr. 1970, ApJ, 159, 379
Schaefer, B. E. 2010, ApJS, 187, 275
Shafter, A. W., & Irby, B. K. 2001, ApJ, 563, 749
Shara, M. M. 1981, ApJ, 243, 926
Strope, R. J., Schaefer, B. E., & Henden, A. A. 2010, AJ, 140, 34
Tang, S., Bildsten, L., Wolf, W. M., et al. 2014, ApJ, 786, 61
Tisserand, P., Le Guillou, L., Afonso, C., et al. 2007, A&A, 469, 387
Tomaney, A. B., & Crotts, A. P. S. 1996, AJ, 112, 2872
Tsapras, Y., Carr, B. J., Weston, M. J., et al. 2010, MNRAS, 404, 604
Udalski, A., Szymanski, M., Kaluzny, J., et al. 1993, AcA, 43, 289
Walterbos, R. A. M., & Kennicutt, R. C. Jr. 1987, A&AS, 69, 311
Williams, S. C., Darnley, M. J., Bode, M. F., Keen, A., & Shafter, A. W. 2014, ApJS, 213, 10
Wyrzykowski, Ł., Kozłowski, S., Skowron, J., et al. 2010, MNRAS, 407, 189
Wyrzykowski, Ł., Kozłowski, S., Skowron, J., et al. 2011a, MNRAS, 413, 493
Wyrzykowski, L., Skowron, J., Kozłowski, S., et al. 2011b, MNRAS, 416, 2949
Zwicky, F. 1933, AcHPh, 6, 110

Chien-Hsiu Lee

Chapter 5

Multi-Wavelength Studies

While high-cadence and long-term monitoring can provide great coverage of the light curves that enable us to study the variability and transients in detail, another essential piece of information in the time-domain studies is panchromatic information. This piece of information is especially important in two aspects: first, we need colors to better characterize the variables, for example, to derive the extinction free Wesenheit magnitude for cleaner Cepheid period–luminosity relations. Second, we might have caught the progenitor of the transients, say novae before outburst, or recurrent novae in the quiescent state. With multi-wavelength observations, we will be able to better pin down the nature of the secondary companion to the central white dwarf. All of these are not possible without panchromatic information. While it is sufficient to have static sky observations rather than multi-epoch multi-wavelength observations to serve for the purpose of characterizing the variables and transients, it is important to note that at the distance of M31, we will need deep and dedicated static sky observations. We will also need to mosaic several pointings even with wide-field cameras, to have a complete picture of the entire M31 galaxy.

Luckily, the philosophy of astronomy is to open and share the data, so we can take advantage of dedicated imaging and spectroscopic surveys to provide complementary color information to the time-domain studies. In this chapter we will give a brief summary of the available resources across the electromagnetic spectrum, from X-ray at the highest frequency end, until radio at the longest wavelength end, as well as dedicated spectroscopic surveys targeting stars, stellar clusters, and planetary nebulae in M31.

5.1 X-Ray

There have been deep X-ray observations of M31 conducted with state-of-the-art space telescopes with the original goal to search for supernovae remnant and X-ray binaries. For example, Kong et al. (2002) have conducted a deep X-ray imaging survey of the central part of M31 with the *Chandra* X-ray observatory and

doi:10.1088/2514-3433/ab3639ch5

discovered globular clusters, supernovae remnants, planetary nebulae, and so forth. There also have been *XMM–Newton* surveys of the northern disk of M31 by Trudolyubov et al. (2005), which uncovered dozens of X-ray bright sources. These X-ray bright sources, when correlated with observations from other wavelengths, turn out to be hot diffuse gas and late-type stars. The most extensive X-ray monitoring of M31 comes from dedicated *XMM–Newton* surveys (see Figure 5.1). For example, Stiele et al. (2011) presented their *XMM–Newton* EPIC M31 survey, with data collected from 2000 to 2008, covering the entire D25 ellipse of M31. They have charted ~1900 X-ray sources, including X-ray binaries, supernova remnants, globular clusters, star-forming regions, and so on. Recently, Stiele & Kong (2018) further employed NuSTAR to study X-ray point sources with energy higher than 10 keV, albeit limited to the central region of M31.

Besides the aforementioned X-ray sources, there is one special class of X-ray sources, called super-soft sources (often dubbed "SSS"), that are mostly associated with novae (see e.g., Pietsch et al. 2007). The SSS-phase is essential in our understanding of novae, especially because the onset and cessation of the SSS-phase can provide insight to the mass of the central white-dwarf (Hachisu & Kato 2006). As novae are transient phenomena and X-ray observation resources are limited from space observatories, it is important to plan X-ray observations efficiently. Luckily, the SSS phase in the X-ray happens sometime after the nova outburst in optical, hence this allows us to plan X-ray observations ahead. In this regard, Henze et al. (2010, 2011, 2014) have conducted extensive joint optical and X-ray observations for several seasons to investigate the novae and their outbursts in M31. They even detected a nova outburst in one of the globular cluster in M31 (Henze et al. 2013).

5.2 Ultraviolet

The UV light, like the X-ray, is also blocked by our atmosphere and we need to conduct the UV observations from space. Nevertheless, it is very rewarding, as the UV light tells us about most recent star formation, and also provide strong constraints in the SED fitting of massive stars. In this regard, there has been a dedicated atlas of nearby galaxies conducted by the *GALEX* satellite, including M31 (Gil de Paz et al. 2007). Making use of this deep and wide UV data set, Kang et al. (2009) have investigated the star-forming regions and the most recent star formation epochs in M31. Kang et al. (2009) reported a slight decrease of star formation rate in the recent epochs, with a peak between 10 and 100 Myr, and an average star formation rate of 0.6–0.7 $M_{\odot}$ per year over the past 400 Myr. Unfortunately, *GALEX* ceased its mission in 2012, but luckily we have the *Swift* satellite, which is still in operation, and also provides sharper images in the UV than *GALEX*.[1] Like X-ray, UV observations also provide important clues to the outburst mechanism of novae (Cao et al. 2012). The best UV imaging and spectrograph data come from state-of-the-art space observatories, especially the UVIS channel of the WFC3 camera on-board the *Hubble Space Telescope* still continuously delivering UV

[1] https://www.nasa.gov/mission_pages/swift/bursts/uv_andromeda.html.

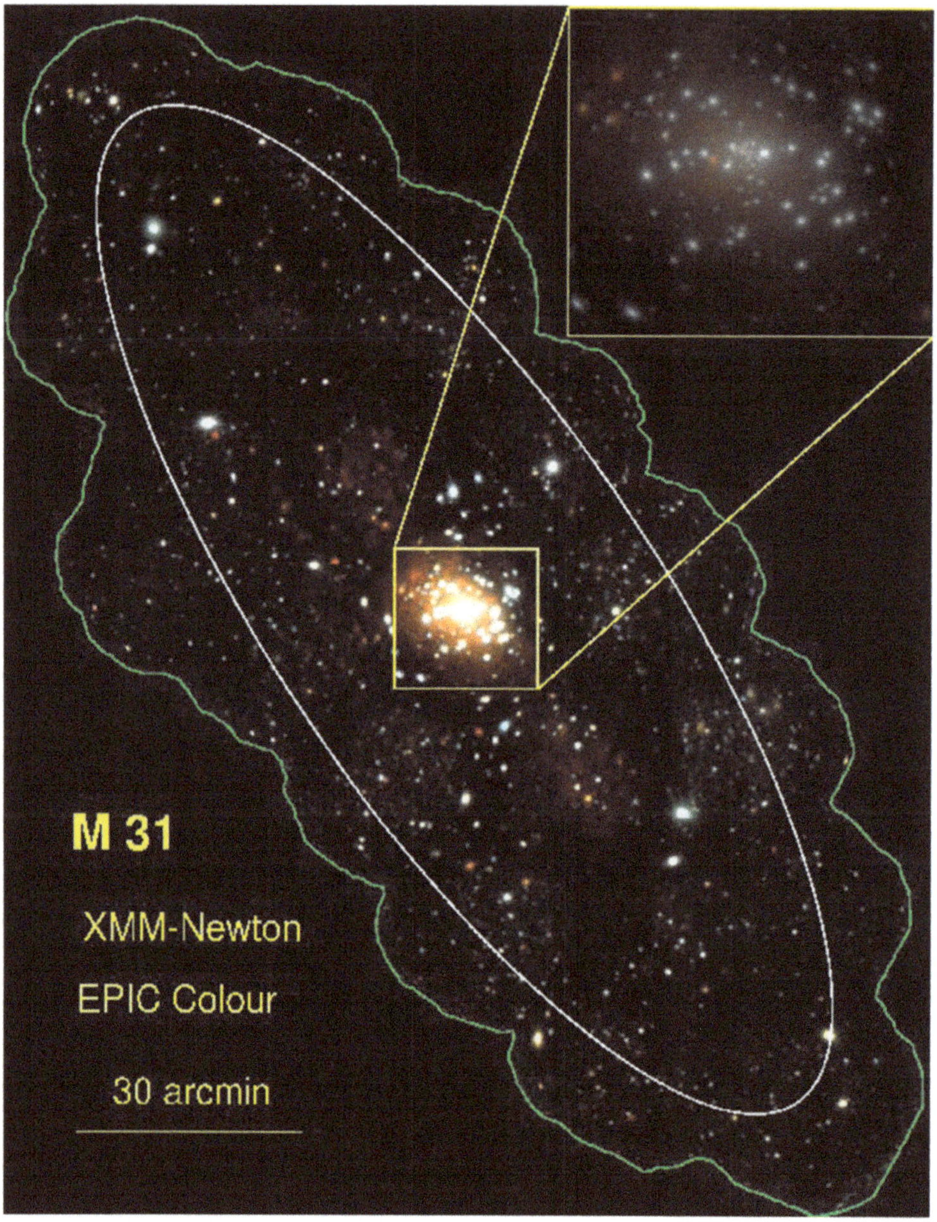

Figure 5.1. X-ray observations of the disk of M31 with the *XMM–Newton* telescope. The colors represent different X-ray energies: 0.2–1 keV in red, 1–2 keV in green, and 2–12 keV in blue. Most of the sources are unidentified hard X-ray sources that could be either X-ray binaries or supernova remnants in M31, or background galaxies. Image credit: Stiele et al. (2011).

imaging at exquisite resolution. For example, Rosenfield et al. (2012) used the WFC3/UVIS with F275W and F336W filters to identify ~4000 UV-bright old stars in the bulge of M31. Such UV-bright populations have been identified before, but their true nature remains debatable. With the exquisite spatial resolution of *HST*, Rosenfield et al. (2012) were able to pin down the nature of these UV-bright objects as extreme horizontal branch stars, and identified the source of UV excess in the old bulge of M31. This exquisite UV imaging data set also provides a good source to examine our knowledge of using the far-UV (FUV) flux as a tracer of recent star formation, as well as the underlying stellar populations. Simones et al. (2014) found that the FUV flux, when correctly corrected for extinction, can provide a reliable star formation rate in recent epochs down to the sub-kpc scale. Please note that to translate the FUV flux into star formation rate we need to have a good handle of initial mass function, and it is under the assumption of a constant star formation rate. In addition, thanks to the exquisite *HST* imaging, Simones et al. (2014) were also able to obtain star formation history (SFHs) of the star-forming regions at a few hundred pc scales, which showed that most of the regions are not from simple stellar populations (SSPs). We should note that SSP is an assumption that is commonly used when inferring star formation rate from integrated FUV flux, especially when the spatial resolution is not sufficient to resolve the star-forming regions in distant galaxies. However, as shown by Simones et al. (2014), a closer look at the star-forming regions with the exquisite spatial resolution of *HST* WFC3/UVIS showed that not all star-forming regions are SSPs; in fact, most of their M31 star-forming regions are not SSPs. In addition, Simones et al. (2014) also showed that when comparing the FUV flux derived star formation estimates to the results from star formation history, they significantly differ in both the age and in the mass.

Another important application of the UV flux is to constrain the extinction curves. Nowadays when dealing with extinction curves, the most commonly used assumption is $R_V = 3.1$. However, as shown by Dong et al. (2014), R_V of 3.1 is not universally true, especially in the center Al region of M31, which exhibits a steeper extinction curve ($R_V \sim 2.5$) that is similar to the findings previously seen in the Galactic bulge, see, e.g., in Nataf et al. (2013).

5.3 Optical

The most extensive deep imaging surveys of M31 come in the optical. In this regard, the Local Group Galaxy Survey conducted by the 4 m Mayall telescope at Kitt Peak National Observatory (Massey et al. 2006) provides the most comprehensive optical view of the entire M31 disk. The Local Group Galaxy Survey identified ~370,000 sources in the U, B, V, R, and I-band filters down to 23 magnitudes with photometric error of <10%, and also employed the narrow-band filters to identify stars emitting strongly in H-alpha, [SII], and [OIII] (Massey et al. 2007). This survey not only provides a detailed map of M31, but also establishes a firm basis for studies of red giants, yellow supergiants, luminous blue variables, and other massive stars. In particular, this survey has identified hundreds of luminous blue variable candidates, which is more than an order of magnitude increase and provides a good sample to study this short-lived phase towards the end stage of massive star evolution. This triggered several follow-up campaigns, especially with spectroscopic observations, and greatly improved our understanding of different evolution stages at the upper HR diagram.

For example, Drout et al. (2009) presented a sample of 54 highly ranked M31 yellow supergiant (YSG) candidates, and 66 probable candidates, among the ~2900 targets they selected as yellow supergiants from color–magnitude diagrams with mass larger than 12 solar mass. This suggests that there is more than 96% of contamination from foreground stars in the Milky Way, which is expected as M31 YSGs are located in a CMD region that is highly contaminated by foreground stars. Nevertheless, with the dozens of YSGs, Drout et al. (2009) showed that the number of YSGs in M31 is an order of magnitude larger than that predicted by theoretical evolutionary tracks (even after taking into account stellar rotations). In addition, they were able to compare the total number of YSGs more massive than 20 solar mass to the total (estimated) number of unevolved O stars to infer how long the YSG phase expands. Surprisingly, they found out that the YSG phase for stars more massive than 20 $M_{\odot}$ lasts only ~3000 years, which is an order of magnitude shorter than the 20,000–80,000 years predicted by evolutionary tracks.

Besides YSGs, Massey et al. (2016) also inspected the M31 red supergiant (RSG) populations using the Local Group Galaxy Survey sample. The RSGs in M31, unlike YSGs, stand out in the B–V versus V-R color–color diagram, and can be easily distinguished from foreground stars in the Milky Way. Massey et al. (2016) then proceeded with spectroscopic follow-up, and were able to determine the spectral type of 192 RSGs out of their 255 candidates. They found out that M31 RSGs show bifurcation distributions in effective temperatures, with one sequence of early K-type RSGs, and another sequence from M-type RSGs. The bifurcation distribution is stronger than predicted from evolution tracks, and can be partially attributed to the metallicity effect. One intriguing finding from Massey et al. (2016) is that RSGs with higher mass spend more time at cooler temperatures.

Beyond YSGs and RSGs, the Wolf-Rayet stars (WRs) are also occupying an important stage of massive star evolution. In particular, the relative number of sub-type WRs, i.e., WC-type and WN-type WRs, depends on the metallicity of the host galaxy. Based on the Local Group Galaxy Survey photometry and further spectroscopic follow-ups, Neugent et al. (2012) identified 107 new WRs, with the majority of them to be WN-type, leading to 154 confirmed M31 WRs at ~95% completeness. The relative number of WC versus WN-type stars and the distribution of WC-type indicate that most of the WRs exist in a high metallicity environment. When compared with evolutionary track predictions, Neugent et al. (2012) found the WC/WN ratio is higher than the theoretical model predictions, suggesting that the rotational velocities in the models could be too high.

While the Local Group Galaxy Survey provides a dedicated map of the disk of M31, it did not go beyond the disk or into the M31 halo. In this regard, the SDSS survey with its larger footprint, albeit much shallower, provides a better overview of different components of M31 (see Figure 5.2). Tamm et al. (2012) utilized the SDSS *ugriz* 5-band optical imaging of M31—combining with the 3.6 μm observations by Spitzer—to perform SED fitting of the light contribution from the entire M31 galaxy and decomposed it into the nucleus, bulge, disk, and halo. They were also able to

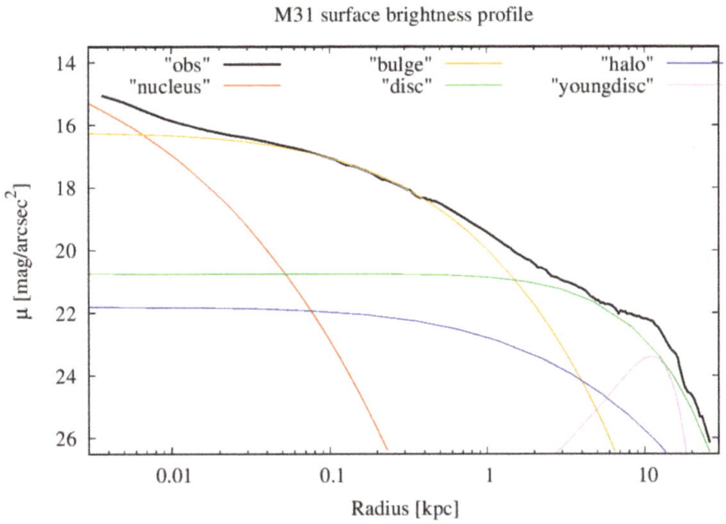

Figure 5.2. Surface brightness profile of M31 as a function of the galactocentric radius along the major axis from SDSS g-band imaging. Data taken from Tempel et al. (2011). The black thick line indicates the observations. Colored lines indicate different components from the best-fit model.

measure the ensemble mass of M31 to be 10–15 × 10^{10} solar mass, and the dark matter viral mass M200 to be 0.8–1.1 × 10^{12} solar mass and the viral radius R200 to be ~200 kpc.

5.4 Infrared

If we want to study the dust properties, or to characterize and avoid extinction effects from the dust, we will need to utilize observations from the infrared (see, e.g., Figure 5.3). This is also true if we want to study the old stellar populations, e.g., RGB and AGB stars, which are bright in the infrared pass-bands. The most extensive near-infrared observations come from the ANDROIDS program (Sick et al. 2014), which uses the WIRCam at the CFHT telescope to provide deep J and K-band imaging of the entire disk of M31, down to a surface brightness level of 16 and 14 mag arcsec^{-2}, respectively. This data set also provides tremendous leverage of the surface brightness profile of M31, and the resolved stellar populations, e.g., RGB stars, that reside within M31. However, there has not been a point-source catalog publicly released yet.

One difficulty of conducting infrared observations from the ground is the fast varying sky brightness due to the atmospheric influences. Indeed, Sick et al. (2014) found it difficult to improve the sky subtraction procedure to better than 2% level even with their sophisticated object-sky nodding observation strategy. In this regard, it would be very beneficial to conduct infrared observations with space-based telescopes, where there are no atmospheric influences on the sky background. One of the most detailed mapping of the M31 disk comes from the extensive *Spitzer* observations (see, e.g., Barmby et al. 2006, Rafiei Ravandi et al. 2016), including the

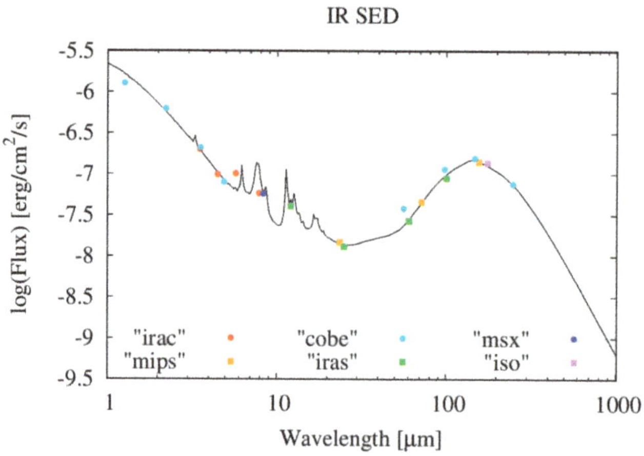

Figure 5.3. Infrared spectral energy distribution of M31. The black line indicates the best-fit model according to Drain & Li (2007). Observations from different space-based probes are shown as follows: *Spitzer IRAC* in red, *Spitzer MIPS* in orange, *COBE* satellite in cyan, *IRAS* satellite in green, *MSX* satellite in blue, and *ISO* satellite in purple. Plotted with data taken from Montalto et al. (2009).

IRAC 3.6, 4.5, 5.8, and 8 μm mosaics presented in (Mould et al. 2008), and the MIPS 24 μm mosaics from Gordon et al. (2006). Making use of these mosaics, Khan (2017) delivered an *Spitzer* point-source catalog of ~860,000 sources in M31. Besides *Spitzer*, the *WISE* spacecraft (Wright et al. 2010) also provides maps of M31 in similar pass-bands (3.4, 4.6, 12, and 22 μm), but with lower spatial resolution and shallower depth (see Figure 5.4). These exquisite space-based IR data sets provide us several insights to M31 properties. First of all, using the mosaic Spitzer 3.6μm images, Seigar et al. (2008) were able to derive a viral mass of $8.2 \times 10^{11}\ M_{\odot}$, in line with dynamical studies using velocities of satellites ($\sim 1.4 \times 10^{12}\ M_{\odot}$; Watkins et al. 2010) and HI observations ($\sim 1.2 \times 10^{12}\ M_{\odot}$; Corbelli et al. 2010). In addition, the 8μm non-stellar emission (L_8), as dominated by the PAH, can also be used to infer the star formation rate. Barmby et al. (2006) used their exquisite and extended *Spitzer* MIR image of M31 and employing a L_8–SFR relation by Wu et al. (2005) to infer an SFR of 0.4 $M_{\odot}$/Myr for M31, in good agreement with SFRs derived from Hα and from far-infrared luminosity. In addition, infrared observations are important to understand the dust properties and extinction of M31 (see, e.g., Montalto et al. 2009). Beyond the properties of M31 as a galaxy, the mid-IR photometry is also valuable for studies of, but not limited to, evolved stars (e.g., Boyer et al. 2015) and testing population synthesis of globular clusters (Barmby & Jalilian 2012).

In the far-infrared end, the most extensive maps come from the Herschel space telescope, specifically from the Herschel exploitation of local galaxy Andromeda (HELGA; Viaene et al. 2014). While there is no source catalog, Viaene et al. did make their Herschel PACS (100, 160 μm) and SPIRE (250, 350, and 500 μm) maps public, enabling interested readers to extract photometry on their own. Making use of the Herschel observations (in combination with *Spitzer* data), Draine et al. (2014)

Figure 5.4. A Panchromatic view of M31. Top: composite three-color RGB UV image from *GALEX*, where the FUV channel is shown in blue, NUV channel is shown in red, and the green comes from the mean of FUV and NUV. The spiral arm is clearly seen in the UV imaging, indicating active ongoing star formation process. Middle: optical three-color image from SDSS using the g, r, and i-band shown as blue, green, and red. The dust lane can be seen as the dark shadow in the optical disk. Bottom: three-color mid infrared image from WISE using the W1 (blue), W2 (green), and W3 (red) channel. The W3 channel is at 12 μm and traces the dust emission. As can be seen, there are copious amounts of dust in the spiral arms.

were able to derive the total dust mass of M31 to be $5.4 \times 10^7\ M_{\odot}$ and a global dust/gas ratio of 0.008 1, though it varies from 0.026 at the center to 0.002 7 at radial distace of 25 kpc.

5.5 Radio

There were many radio telescope observations of M31, with the goals to study molecular gases, to characterize the rotation curves (see, e.g., Figure 5.5) and to infer the total mass of the neutral gas M31. For example, Nieten et al. (2006) used the IRAM 30 m telescope to obtain CO observations of M31. Using a constant X_{CO} value, they were able to obtain a total molecular mass of 3.6×10^8 $M_\odot$ within 18 kpc. This, when combined with the HI mass, indicates that the total neutral gas in M31 is 5.2×10^9 $M_\odot$.

Carignan et al. (2006) used the Effelsberg and Green Bank 100 m telescopes to map the rotation curve of M31 out to a radial distance of 35 kpc using HI observations. They showed that the rotation curve is nearly flat at 226 km s^{-1} between 20 and 35 kpc, and derived a total mass of 3.4×10^{11} $M_\odot$ at projected distance <35 kpc. Furthermore, Chemin et al. (2009) used the Synthesis Telescope and the 26 m Dominion Radio Astrophysical Observatory and presented an updated 21 cm map of M31. Their results indicated new HI structures, including an external arm and thin HI spurs in the outskirts of the disk, and suggested that these features are the results of tidal interactions between M31 and its companion galaxies.

Using the Very Large Array, Galvin et al. (2012) presented a high resolution (<10'), high sensitivity (60 μ Jy) 20 cm map of M31. The most prominent feature in this wavelength is the 10 kpc star-forming ring, which they estimated a total integrated flux of 706 mJy. They also reported 864 unique discrete radio sources,

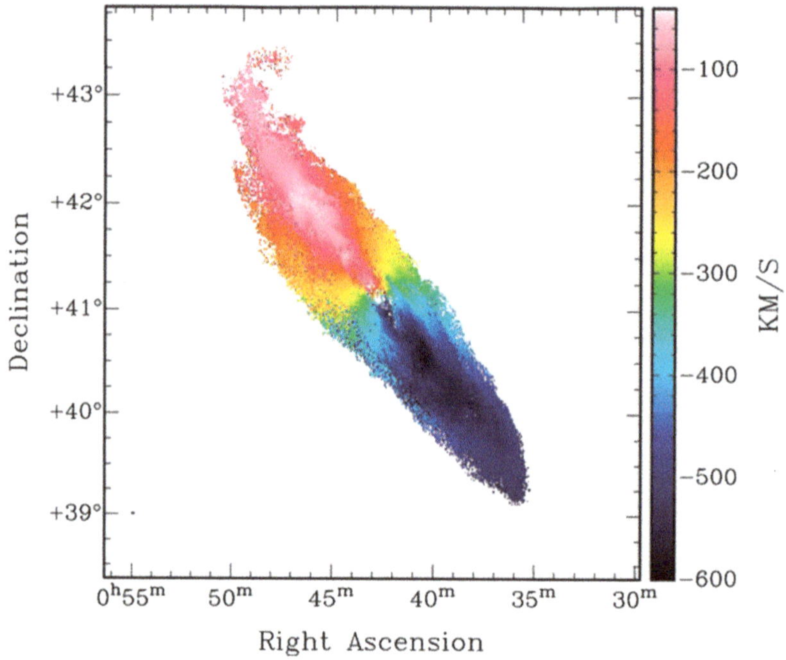

Figure 5.5. Velocity field of M31 from 21 cm HI observations. Image credit: Chemin et al. (2009)

however, with the majority of them being non-thermal and are more likely to be background objects.

5.6 Pan-Andromeda Archaeological Survey

As the closest spiral galaxy, M31 is also an ideal testbed for galaxy formation and interactions. There have been discoveries of tidal streams in the halo of M31 by deep imaging and RGB star counting (Ibata et al. 2001). To have a better view of the tidal streams and substructures in the halo of M31 (and also extending to M33), the Pan-Andromeda Archaeological Survey (PAndAS) team used the wide imaging and deep photometry enabled by the 1 squared degrees camera MegaCAM mounted on the 3.6 m Canada–France–Hawaii Telescope to deliver a panoramic view of the M31 halo in the SDSS g and i-filters (McConnachie et al. 2009). The footprint of PAndAS covers ~400 squared degrees, mainly the M31 halo out to a projected radius of 150 kpc, and also extends to and covers M33 and its halo out to a projected radius of 50 kpc (see Figure 5.6).

The PAndAS survey has expanded our understanding of substructures and satellite galaxies from M32, NGC 205, NGC 147, NGC 158 to beyond dozens of dwarf galaxies, tidal structures, global clusters, and so forth. For the dwarf galaxies, the main hurdle is to distinguish the RGBs in the halo of M31 from contaminations by foreground stars sitting in the Milky Way. In this regard, Martin et al. (2013) have devised a Bayesian approach that can distinguish a foreground contamination from a RGB star in the M31 halo. The idea is, given the information of the position and the g-i color of a source towards the M31 line-of-sight, we can estimate the

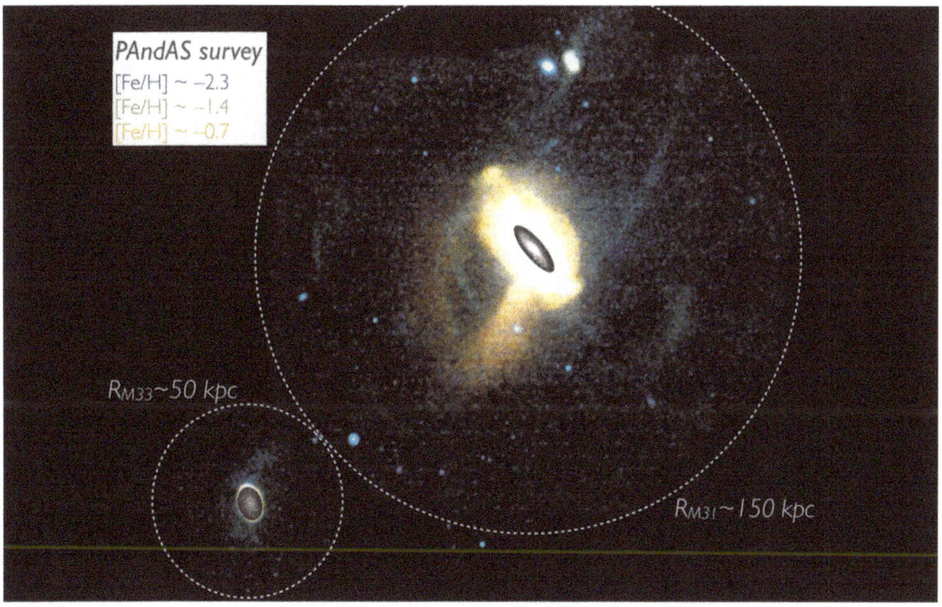

Figure 5.6. Giant stellar stream and tidal features in the halo of M31 as identified by the PAndAS survey. Image credit: Martin et al. (2013) © 2013. The American Astronomical Society. All rights reserved.

likelihood of this source being a foreground star that matches the Milky Way model, or an intrinsic RGB star in the M31 halo. Comparing the likelihood of both scenarios can then be used to discard foreground contaminating sources. Following up on the list of candidate dwarf galaxies using this approach, Martin et al. (2016) presented detail properties of 23 M31 dSphs. Also, with the dwarf galaxies discovered by PAndAS, Ibata et al. (2013) gave a close look of 27 dwarf galaxies, and found out that 15 of the dwarf satellite galaxies lie in a vast thin plane orbiting M31. More strikingly, 13 out of the 15 satellite galaxies share the same rotational motion. Such co-planar motion of satellite galaxies is rarely seen in cosmological simulations, hinting to caveats in the ΛCDM, or there might be missing physical inputs in the cosmological simulations.

The PAndAS data also enable scrutiny of tidal features in M31. For example, Conn et al. (2016) used the PAndAS data and revised our understanding of the giant stellar stream. First of all, they confirmed a distance gradient of 20 kpc/degree, with the farthest part of the giant stellar stream to be 120 kpc more distant than the closest part of the stream. Second, the metallicity gradient of the giant stream appears to be piece-wise. It steadily increases from [Fe/H] = −0.7 to −0.2 along the inner part of the stream, and then steadily drops to [Fe/H] = −1.0 at the farthest edges of the PAndAS coverage. Another major feature in the halo of M31 is the South-West Cloud. Using the PAndAS data, Bate et al. (2014) studied this substructure in detail of the first time, derived a distance of 793 ± 45 kpc, a metallicity of [Fe/H] = −1.3, and a brightness of $M_V = -12.1$ mag; all of these suggest that the progenitor of the South-West Cloud is among the brightest dwarf galaxies prior to accretion by M31.

The PAndAS survey is also a great resource for the globular cluster studies. For example, Huxor et al. (2011, 2014) have investigated the globular clusters in the halo of M31 using the exquisite PAndAS imaging. They found 59 new bona fide globular clusters, and when examining the candidates reported by the Revised Bologna Catalogue (RBC) and from SDSS, 188 previously reported candidates are either foreground stars or background galaxies. The PAndAS data can reach M31 globular clusters that are as faint as $M_V = -6$ mag, and with artificial cluster tests Huxor et al. estimated a 50% completeness at $M_V = -4.1$ mag. Using this exquisite data set, Huxor et al. (2014) examined the luminosity function of M31 globular clusters, and found that it is bimodal in the outer halo (projected radius >30 kpc), with a second peak at $M_V = -5.5$. Huxor et al. suggested that this second peak represents globular clusters that were accreted along with their host dwarf galaxies by M31.

5.7 Panchromatic Hubble Andromeda Treasury

Besides the aforementioned surveys of the entire disk of M31, there is also a dedicated Hubble Space Telescope program—Panchromatic Hubble Andromeda Treasury (dubbed as PHAT)—that makes use of the exquisite spatial resolution of the ACS and WFC3 instruments on-board *HST*, to map the northern 1/3 of the M31 disk (see Figure 5.7) in great detail in UV, optical, and near-infrared pass-bands

Figure 5.7. Mosaics of the PHAT observations in the optical pass-band. Image credit: NASA, ESA, J. Dalcanton (University of Washington, USA), B. F. Williams (University of Washington, USA), L. C. Johnson (University of Washington, USA), the PHAT team, and R. Gendler.

(Dalcanton et al. 2012). PHAT aims at limiting magnitudes (4-sigma) down to 25 magnitudes in the UV, 27 magnitudes in the optical (F475W and F814W, corresponding to g- and i-band, respectively), and 25 magnitude in the near-infrared (F110W and F160W, corresponding to J and H-band) in the uncrowded parts of the disk.

This exquisite data set not only allows us to have a sharp view of the M31 disk by resolving the disk into individual stars, but also provides a firm basis to study stellar clusters, initial mass functions, star formation histories, dust properties, and extinction law in M31 in detail. The observations were carried out in parallel mode using both ACS and WFC3 (UVIS and IR imager). Since the WFC3 detectors have smaller fields-of-view, the ACS parallel imaging were covering much larger (and overlapping) fields in the optical pass-bands. In this regard, the PHAT data also contain multiple epochs, from two epochs in the UVIS and IR, to a handful of epochs in the optical, which can also be used to derive sparse time-series photometry of variables. This is very helpful to identify variables when compared to ground-based observations where there might be hundreds of stars in one pixel of the ground-based imager(s). The high level product, including dedicated PSF photometry and mosaicked images, are publicly available at the STScI website.[2]

The detail of the PHAT photometry was presented in Williams et al. (2014), composed of 117 resolved stars in 6 *HST* filters from UV to NIR (F275W, F336W, F475W, F814W, F110W, and F160W). The exquisite photometry enables us to resolve star clusters even at the distance of M31 (see, e.g., Fouesneau et al. 2014, Johnson et al. 2015). In particular, Johnson et al. (2015) presented 2573 stellar clusters and 2270 background galaxies visually inspected by citizen scientists (the Andromeda Project) and estimated the completeness to be 50% down to clusters with $\sim$500 $M_{\odot}$ and ages <100 Myr. With this sample, Weisz et al. (2015) were able to examine the initial mass function (IMF) at the high mass end (M > 2 $M_{\odot}$). They found the IMF has a slope of $\Gamma = 1.45^{+0.03}_{-0.06}$, which is steeper than the Kroupa (1.3) and Salpeter (1.35) values, with significant improvements in precision. In addition, the Γ value of M31 is more precise and differs from the Γ of the Milky Way (1.15 $\pm$ 0.01) and the Large Magellanic Cloud (1.3 $\pm$ 0.1). When combining the high mass end of IMF from their studies, and the widely used sub-solar mass Kroupa and Chabrier IMF, Weisz et al. found that the commonly used UV and Hα indicators for star formation rate estimates are underestimated by a factor of 1.3–1.5. In addition, the number of stars with masses >8 $M_{\odot}$ is 25% fewer than expected from the Kroupa and Salpeter IMF.

The deep PHAT photometry also enables studies of star formation history using color–magnitude diagrams (CMDs). In this regard, Lewis et al. (2015) were able to study the star formation history within the last 500 Myr. Their findings showed that the 10 kpc ring is at least 400 Myr old, with ongoing star formation activity. Most importantly, the 10 kpc ring was stationary throughout this period, almost ruling out a collisional origin (i.e., M31 head-on collision with M32 or small galaxies).

[2] https://archive.stsci.edu/prepds/phat/.

They also obtained a star formation rate of ~0.7 $M_{\odot}yr^{-1}$ for the entire M31 disk, which is consistent with broadband photometry of the entire M31 disk. Tracing star formation history using CMD is more difficult, nevertheless thanks to the exquisite PHAT photometry, Williams et al. (2017) were able to determine the star formation history for the past 14 Gyr. First of all, they found most of the star formation happened >8 Gyr ago, and then became quiescent until 4 Gyr ago. The next star formation episode happened 2 Gyr ago, then became quiescent again. However, there are no clear structures with stars older than 2 Gyr, indicating a significant mixing after the 2 Gyr star formation episode.

The PHAT photometry, especially in the infrared, also provides a great leverage to a high resolution dust extinction map. Dalcanton et al. (2015), making use of the near-infrared CMD of RGBs, were able to obtain a sharp extinction map, which is a factor of four better than the maps from dust emission by Draine et al. (2014). More importantly, this sharp PHAT dust map yields a direct measurement of the dust column in M31. When compared with the commonly employed dust models by Draine & Li (2007), Dalcanton et al. (2015) found that the models over-predicted the extinction by a factor of 2.5, and suggested that Draine et al. (2014) overestimated the total dust mass in M31.

5.8 Spectroscopic Surveys

When it comes to multi-wavelength information, nothing works better than taking spectra of the targets of interests. The many deep and wide imaging surveys of M31 provide a great list of intriguing sources that warrant spectroscopic follow-up. The only down side is that as M31 is very distant, we will need very long integration time (even with large aperture telescopes) to collect spectra with reasonable S/N. In addition, as M31 is very extended, we also need multi-object spectrograph that can cover a wide area, to obtain spectra for a large number of M31 sources in a more effective and efficient manner. In this regard, the SPLASH survey (Guhathakurta et al. 2006) has been gathering spectra of M31 stars (mostly RGBs) using the multiplexing spectrograph DEIMOS on the Keck 10 m telescope. The RSGs are evolved stars and can be easily identified in the halo of M31 using the intermediate band filter DDO51. Since the RSGs are old, they can trace tidal features due to interactions of M31 with other galaxies, and provide insights to the formation of M31. The most spectacular tidal feature is the giant southern stream, discovered by Ibata et al. (2001). Thanks to the generous field-of-view of DEIMOS, Guhathakurta et al. (2006) were able to obtain spectra for 68 bona fide M31 RGBs, and determined their radial velocity and relative higher metallicity compared to other parts of the M31 spheroid, and in turn inferred the nature of the progenitor of the stream to be a luminous dwarf galaxy. Further Keck/DEIMOS follow-up showed that the giant southern stream extended into the south eastern inner halo of M31 (Gilbert et al. 2007, 2009), matching the predictions from N-body simulations of interactions between M31 and the progenitor of the giant southern stream from Fardal et al. (2006, 2007). Using this unique sample of RGBs, Gilbert et al. (2012, 2014) were able to constrain the surface brightness and metallicity profile of M31 out to a projected distance of 175 kpc. They observed strong field-to-field variation in surface brightness and metallicity beyond 90 kpc, indicating that the outermost halo of M31 exhibits evidence that multiple smaller progenitors have contributed to the halo stars in M31. In addition, taking advantage of the vast amount of satellite galaxies identified by the CFHT imaging, Tollerud et al. (2012) presented a systematic view of 15 M31 dwarf Spheroidal satellites (dSphs), to obtain their kinematics, mass, and mass–size–luminosity relations. Their results suggest that the properties of the M31 dSphs are consistent with those in the Milky Way dSphs.

In addition to SPLASH, there is also another spectroscopy survey using the multiplexing power of the Hectospec instrument at the 6.5 m MMT[3] (Caldwell et al. 2009). The MMT survey obtained spectra for stellar clusters, planetary nebulae, HII regions, and stars in M31. For example, Sanders et al. (2012) identified 253 HII regions and 407 planetary nebulae with Hectospec/MMT, and used them to derive the metallicity gradient of M31, which shows a shallow oxygen abundance gradient, consistent with the result of Zaritsky et al. (1994). However, they also demonstrated that oxygen abundance directly extracted from the electron temperature differs significantly from that of the strong-line ratio method. Furthermore, it is also possible to obtain spectra of bright stars in M31 even using 4 m class telescopes. For example, the Local Group Galaxy Survey (Massey et al. 2007) made use of the multiplexing power provided by the Hydra spectrograph mounted on the WIYN telescope, providing follow-up spectra of intriguing emission line objects, e.g., LBVs, Wolf-Rayet stars, HII regions, and so forth, that were pre-selected from the Hα, OIII, and SII narrow-band filter imaging (Massey et al. 2007). The deep spectra from the SPLASH, MMT, and the Local Group Galaxy surveys provide unique opportunities to study these objects beyond the Milky Way.

5.9 Relation to Time-Domain Studies

The aforementioned multi-wavelength data sets are related to the time-domain studies in two aspects. First of all, while the vast majority of the time-domain studies are conducted in the optical wavelength, time-series observations in other wavelengths are also important to shed light on the nature of the transients and variables. For example, the super-soft X-ray phase of a nova outburst, especially the onset and cessation of this phase, can provide important information to estimate the mass of the central white dwarf. As for variable stars, the shape of their light curves may appear quite different in different wavelength. For example, in the UV RR Lyra stars exhibit light curve amplitudes about twice as large as in the V-band. This makes it very useful to identify RR Lyra stars, especially the low amplitude ones, in the UV (see, e.g., Kinman & Brown 2014). When it comes to longer wavelength, especially in the near and mid infrared, the light curve amplitudes of RR Lyra stars become really small (see, e.g., Klein et al. 2011). While this is a disadvantage to identify RR Lyra stars, it is actually a great advantage to use RR Lyra stars as distance indicators. This is because their light curve amplitudes are so small, such that we can use random phased

[3] https://http://www.cfa.harvard.edu/oir/eg/m31clusters/M31_Hectospec.html.

photometry, combined with period determined from the optical or UV, to establish a clean period–luminosity relation (see Klein et al. 2014). This is the same case for the Cepheids. Indeed, Riess et al. (2012) and Kodric et al. (2015) have coupled the periods of M31 Cepheids from ground-based observations with random phased photometry from the *Hubble Space Telescope*, to derive a very tight infrared period–luminosity relation (see Section 3.1 in this book). Another merit of infrared observations is that the reddening effect is relatively small comparing to the observations in the optical. This means that the reddening is minimized if we employed infrared observations. Indeed, the infrared period–luminosity relation of Cepheids in M31 presented in Riess et al. (2012) is very tight even without using the Wesenheit magnitude. In fact, infrared observations are essential to tackle the reddening effect, or the dust extinction issue. For example, in Bonanos et al. (2006) the infrared photometry plays an essential role in determining the $E(B - V)$ and R_V such that it is possible to infer distance to M33 with an eclipsing binary system. As a matter of fact, Vilardell et al. (2010) mentioned that about half of the uncertainty of the M31 eclipsing binary distance estimates origin from the unconstrained reddening effect. If we can employ the exquisite infrared photometry from PHAT, we will be able to tackle this issue. In addition, infrared observations also provide an important avenue to discover transients that are highly obscured by the dust with faint or even no optical counterparts, like the eSPecially Red Intermediate Luminosity Transient Events (SPRITEs, Kasliwal et al. 2017). The radio observations might be less directly related to the transient and varying sources residing in M31. This is most of the M31 radio transients and variables will be too faint to be reached with current facilities, and most of the radio detections will be either in foreground sources in the Milky Way, or from background galaxies. Nevertheless, the radio observations are essential to rule out those two contaminations (Table 5.1). For example, radio observations are useful to rule out quasars that are in the background of M31 but misclassified from optical light curves alone (see, e.g., Massey et al. 2019).

Table 5.1. Multi-wavelength Observations, Related Properties, and Corresponding Variables.

Observations	Properties	Corresponding variables
X-ray	SN, X-ray binaries, super-soft X-rays	eclipsing binaries, novae
UV	Recent star formation	Cepheids
Hα	Current star formation	Cepheids, LBVs
Optical	Star formation in the past Gyr	Cepheids, LPVs
Near infrared	Star formation across galaxy lifetime	Cepheids, LPVs, RR Lyra stars
Mid and far-infrared	PAHs and dusts	LPVs (dust production), RR Lyra stars (extinction estimate)
Radio	molecular clouds and star-forming clouds	massive stars (such as LBVs)

Table 5.2. Basic Properties of Milky Way and the Andromeda Galaxy.

Property	Milky Way	M31
	Global properties	
Luminosity		
L_V ($10^{10} L_{V\odot}$)	2.1	2.6–2.7
L_K ($10^{10} L_{K\odot}$)	5.5	6–12
Color (B–V)	0.84	0.81
Mass (10^{10} $M_\odot$)	40–55	107–140
	Bulge	
Mass (10^{10} $M_\odot$)	1–2	3.2
Radius (kpc)	2.5	2.6
	Disk	
Scale length		
in V-band	2.5–3.5	6
in K-band	2.3–2.8	4.8
SFR ($M_\odot$/yr)	1–5	0.3–1
Mass (10^{10} $M_\odot$)	3.5	7
Stellar mass (10^{10} $M_\odot$)	3	6
Gas mass (10^{10} $M_\odot$)	0.7	0.6
HI mass (10^{10} $M_\odot$)	0.4	0.5
H_2 mass	0.11	0.02–0.04
Metallicity gradient [O/H] (dex kpc^{-1})	−0.04 to −0.07	−0.018 to −0.027

The multi-wavelength observations also provide a comprehensive picture of the formation and evolution of the Andromeda galaxy, not only in the merging/interaction of companion galaxies, but also in details of the star formation history, dust, and gas contents. Here we provide a brief summary of how multi-wavelength studies are pertinent to the properties of the Andromeda galaxy, such as star formation history, dust content, etc. as well as the corresponding variables or transients that can be used for the same studies or putting constraints.

From these dedicated multi-wavelength observations, we were able to gain a comprehensive picture of the properties of M31, and it would be interesting to compare these properties to that from the Milky Way. Such kind of comparison has been done, e.g., by Yin et al. (2009). Here we summarize some of the basic properties comparing the Milky Way and M31 in Table 5.2.

References

Barmby, P., Ashby, M. L. N., Bianchi, L., et al. 2006, ApJ, 650, L45
Barmby, P., & Jalilian, F. F. 2012, AJ, 143, 87
Bate, N. F., Conn, A. R., McMonigal, B., et al. 2014, MNRAS, 437, 3362
Bonanos, A. Z., Stanek, K. Z., Kudritzki, R. P., et al. 2006, ApJ, 652, 313
Boyer, M. L., McQuinn, K. B. W., Barmby, P., et al. 2015, ApJS, 216, 10

Caldwell, N., Harding, P., Morrison, H., et al. 2009, AJ, 137, 94
Cao, Y., Kasliwal, M. M., Neill, J. D., et al. 2012, ApJ, 752, 133
Carignan, C., Chemin, L., Huchtmeier, W. K., & Lockman, F. J. 2006, ApJ, 641, L109
Chemin, L., Carignan, C., & Foster, T. 2009, ApJ, 705, 1395
Conn, A. R., McMonigal, B., Bate, N. F., et al. 2016, MNRAS, 458, 3282
Corbelli, E., Lorenzoni, S., Walterbos, R., Braun, R., & Thilker, D. 2010, A&A, 511, A89
Dalcanton, J. J., Fouesneau, M., Hogg, D. W., et al. 2015, ApJ, 814, 3
Dalcanton, J. J., Williams, B. F., Lang, D., et al. 2012, ApJS, 200, 18
Dong, H., Li, Z., Wang, Q. D., et al. 2014, ApJ, 785, 136
Draine, B. T., Aniano, G., Krause, O., et al. 2014, ApJ, 780, 172
Draine, B. T., & Li, A. 2007, ApJ, 657, 810
Draine, B. T., Aniano, G., Krause, O., et al. 2014, ApJ, 780, 172
Drout, M. R., Massey, P., Meynet, G., Tokarz, S., & Caldwell, N. 2009, ApJ, 703, 441
Fardal, M. A., Babul, A., Geehan, J. J., & Guhathakurta, P. 2006, MNRAS, 366, 1012
Fardal, M. A., Guhathakurta, P., Babul, A., & McConnachie, A. W. 2007, MNRAS, 380, 15
Fouesneau, M., Johnson, L. C., Weisz, D. R., et al. 2014, ApJ, 786, 117
Galvin, T. J., Filipovic, M. D., Crawford, E. J., et al. 2012, SerAJ, 184, 41
Gilbert, K. M., Fardal, M., Kalirai, J. S., et al. 2007, ApJ, 668, 245
Gilbert, K. M., Guhathakurta, P., Kollipara, P., et al. 2009, ApJ, 705, 1275
Gilbert, K. M., Guhathakurta, P., Beaton, R. L., et al. 2012, ApJ, 760, 76
Gilbert, K. M., Kalirai, J. S., Guhathakurta, P., et al. 2014, ApJ, 796, 76
Gil de Paz, A., Boissier, S., Madore, B. F., et al. 2007, ApJS, 173, 185
Gordon, K. D., Bailin, J., Engelbracht, C. W., et al. 2006, ApJ Letter, 638, L87
Guhathakurta, P., Rich, R. M., Reitzel, D. B., et al. 2006, AJ, 131, 2497
Hachisu, I., & Kato, M. 2006, ApJS, 167, 59
Henze, M., Pietsch, W., Haberl, F., et al. 2010, A&A, 523, A89
Henze, M., Pietsch, W., Haberl, F., et al. 2011, A&A, 533, A52
Henze, M., Pietsch, W., Haberl, F., et al. 2013, A&A, 549, A120
Henze, M., Pietsch, W., Haberl, F., et al. 2014, A&A, 563, A2
Huxor, A. P., Ferguson, A. M. N., Tanvir, N. R., et al. 2011, MNRAS, 414, 770
Huxor, A. P., Mackey, A. D., Ferguson, A. M. N., et al. 2014, MNRAS, 442, 2165
Ibata, R., Irwin, M., Lewis, G., Ferguson, A. M. N., & Tanvir, N. 2001, Nature, 412, 49
Ibata, R. A., Lewis, G. F., Conn, A. R., et al. 2013, Natur, 493, 62
Johnson, L. C., Seth, A. C., Dalcanton, J. J., et al. 2015, ApJ, 802, 127
Kang, Y., Bianchi, L., & Rey, S.-C. 2009, ApJ, 703, 614
Kasliwal, M. M., Bally, J., Masci, F., et al. 2017, ApJ, 839, 88
Khan, R. 2017, ApJS, 228, 5
Kinman, T. D., & Brown, W. R. 2014, AJ, 148, 121
Klein, C. R., Richards, J. W., Butler, N. R., & Bloom, J. S. 2011, ApJ, 738, 185
Klein, C. R., Richards, J. W., Butler, N. R., & Bloom, J. S. 2014, MNRAS, 440, L96
Kodric, M., Riffeser, A., Seitz, S., et al. 2015, ApJ, 799, 144
Kong, A. K. H., Garcia, M. R., Primini, F. A., et al. 2002, ApJ, 577, 738
Lewis, A. R., Dolphin, A. E., Dalcanton, J. J., et al. 2015, ApJ, 805, 183
Martin, N. F., Ibata, R. A., Lewis, G. F., et al. 2016, ApJ, 833, 167
Martin, N. F., Ibata, R. A., McConnachie, A. W., et al. 2013, ApJ, 776, 80
Massey, P., McNeill, R. T., Olsen, K. A. G., et al. 2007, AJ, 134, 2474

Massey, P., Neugent, K. F., & Smart, B. M. 2016, AJ, 152, 62
Massey, P., Neugent, K. F., & Levesque, E. M. 2019, AJ, 157, 227
Massey, P., Olsen, K. A. G., Hodge, P. W., et al. 2006, AJ, 131, 2478
McConnachie, A. W., Irwin, M. J., Ibata, R. A., et al. 2009, Natur, 461, 66
Montalto, M., Seitz, S., Riffeser, A., et al. 2009, A&A, 507, 283
Mould, J., Barmby, P., Gordon, K., et al. 2008, ApJ, 687, 230
Nataf, D. M., Gould, A., Fouqué, P., et al. 2013, ApJ, 769, 88
Neugent, K. F., Massey, P., & Georgy, C. 2012, ApJ, 759, 11
Nieten, C., Neininger, N., Guélin, M., et al. 2006, A&A, 453, 459
Pietsch, W., Haberl, F., Sala, G., et al. 2007, A&A, 465, 375
Rafiei Ravandi, M., Barmby, P., Ashby, M. L. N., et al. 2016, MNRAS, 459, 1403
Riess, A. G., Fliri, J., & Valls-Gabaud, D. 2012, ApJ, 745, 156
Rosenfield, P., Johnson, L. C., Girardi, L., et al. 2012, ApJ, 755, 131
Sanders, N. E., Caldwell, N., McDowell, J., & Harding, P. 2012, ApJ, 758, 133
Seigar, M. S., Barth, A. J., & Bullock, J. S. 2008, MNRAS, 389, 1911
Sick, J., Courteau, S., Cuillandre, J.-C., et al. 2014, AJ, 147, 109
Simones, J. E., Weisz, D. R., Skillman, E. D., et al. 2014, ApJ, 788, 12
Stiele, H., & Kong, A. K. H. 2018, MNRAS, 475, 4911
Stiele, H., Pietsch, W., Haberl, F., et al. 2011, A&A, 534, A55
Tamm, A., Tempel, E., Tenjes, P., Tihhonova, O., & Tuvikene, T. 2012, A&A, 546, A4
Tempel, E., Tuvikene, T., Tamm, A., & Tenjes, P. 2011, A&A, 526, A155
Tollerud, E. J., Beaton, R. L., Geha, M. C., et al. 2012, ApJ, 752, 45
Trudolyubov, S., Kotov, O., Priedhorsky, W., Cordova, F., & Mason, K. 2005, ApJ, 634, 314
Viaene, S., Fritz, J., Baes, M., et al. 2014, A&A, 567, A71
Vilardell, F., Ribas, I., Jordi, C., Fitzpatrick, E. L., & Guinan, E. F. 2010, A&A, 509, A70
Watkins, L. L., Evans, N. W., & An, J. H. 2010, MNRAS, 406, 264
Weisz, D. R., Johnson, L. C., Foreman-Mackey, D., et al. 2015, ApJ, 806, 198
Williams, B. F., Lang, D., Dalcanton, J. J., et al. 2014, ApJS, 215, 9
Williams, B. F., Dolphin, A. E., Dalcanton, J. J., et al. 2017, ApJ, 846, 145
Wright, E. L., Eisenhardt, P. R. M., Mainzer, A. K., et al. 2010, AJ, 140, 1868
Wu, H., Cao, C., Hao, C. N., et al. 2005, ApJ, 632, L79
Yin, J., Hou, J. L., Prantzos, N., et al. 2009, A&A, 505, 497
Zaritsky, D., Kennicutt, R. C., & Huchra, J. P. 1994, ApJ, 420, 87

Chapter 6

Prospects: Deeper, Redder, and Longer

While the archived M31 time-domain data from the Canada–France–Hawaii Telescope, the Isaac Newton Telescope, the Pan-STARRS, the Palomar Transient Factory, and so on already provide a great leverage on bright variable stars with periods above one day—especially the Cepheids that can be as faint as $m_r = 23$ magnitudes—these archival data are not deep enough to reveal the faintest population of variables, especially optical counterparts of the X-ray binaries and the shortest period distance indicator—RR Lyra stars.

The importance of RR Lyra stars cannot be overemphasized. RR Lyra stars are very old (>10 Gyr) and their presence already indicate the existence of old stellar population. Given their old age, they are often used as tracers of tidal streams or substructures to study the interaction and accretion of the Milky Way and its companion galaxies (see, e.g., Figure 6.1). Because of their old age, RR Lyra stars can also provide insight on the Galaxy formation process at the earliest stage. Second, the periods and the amplitudes of the light curves of RR Lyra stars are affected by their metallicity (Sesar et al. 2010); hence, they can also be used as a metallicity tracer, especially for the substructures they reside in. In addition, RR Lyra stars pose a fixed color during the light curve minimum, and we can use this characteristic to determine the line-of-sight extinction (Kunder et al. 2010). With their metallicity in hand, we can use empirical relations to estimate their intrinsic luminosity (Cacciari & Clementini 2003) and in turn derive their distances—with the aforementioned fixed minimum color–in an extinction free manner. Thanks to these advantages, RR Lyra stars provide a unique opportunity to identify tidal streams and substructures in M31 and to characterize the 3D structure and metallicities of these streams and substructures. This is especially the case for streams and substructures across the disk of M31, because there are too many sources in the disk of M31, and the canonical RBG star counting method to search for tidal features is difficult to apply.

doi:10.1088/2514-3433/ab3639ch6

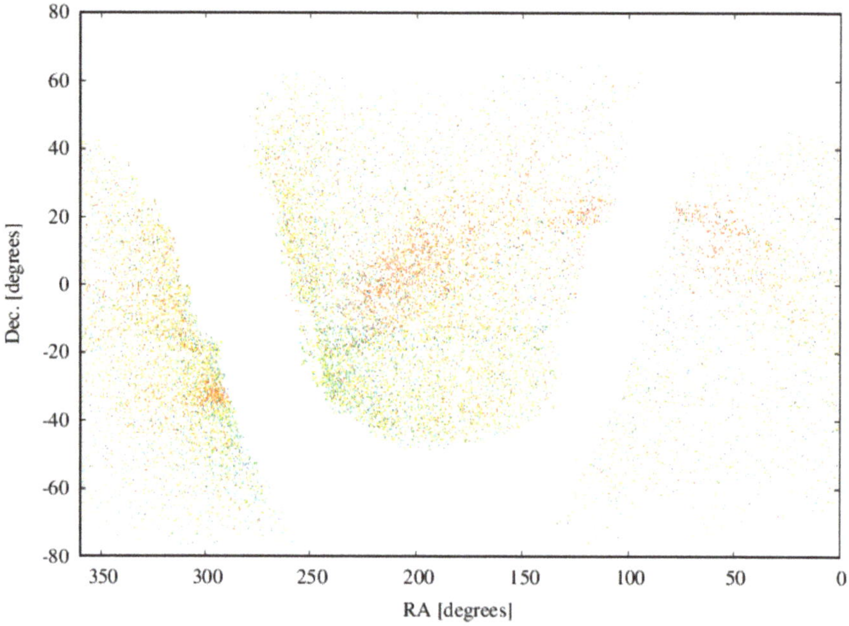

Figure 6.1. Spatial distribution of the RR Lyra (RRab) stars discovered from the Catalina Sky Survey, color coded using their visual magnitudes. The color scheme is as follows: $V > 18$ mag in red, $18 > V > 17$ mag in orange, $17 > V > 16$ mag in yellow, $16 > B > 15$ mag in green, $5 > V > 14$ mag in cyan, and $14 > V > 13$ mag in blue. The Sagittarius stream is clearly seen from the $V > 18$ mag RRab, spanning from (RA, Dec) = (0, 0) to (100, 80) and then to (250, −20). Plotted with data taken from Drake et al. (2013a, 2013b) and Torrealba et al. (2015).

The only disadvantage of RR Lyra stars is that at the distance of M31, they will be very faint, at 25 mag level in the g- and r-band. Due to this fact, the previous searches of RR Lyra stars in M31 were either concentrated on some very small regions of M31 with the *HST* (Sarajedini et al. 2009; Jeffery et al. 2011) or limited to RR Lyra stars with longer periods because of the need of long exposure (at least an hour) from mid-class telescopes (Pritchet & van den Bergh 1987; Dolphin et al. 2004). However, the long exposure will smear out the light curve amplitude and hamper the identification of RR Lyra stars at the short-period end. For example, for RR Lyra stars with period as short as 0.2 days and at the very faint end of the period–luminosity relation, we will need at least an hour of integration with a 4 m telescope to detect them. Please note that an hour of integration will span 20% of the pulsation cycle, significantly smearing out the light curves, and thus preventing us from determining the periods and light curve amplitudes. To obtain decently sampled light curves of RR Lyra stars, we will need 8 m telescopes to reach the desired depth for M31 RR Lyra stars given their short period and faint nature.

However, there are not many wide-field cameras on 8 m telescopes, which is essential to provide a census of the RR Lyra stars population across the disk of M31. Currently, there are only two suitable instruments/telescopes. The first one is the Hyper Suprime-CAM (HSC) mounted on the 8.2 m Subaru telescope. The HSC has a field-of-view of 1.5 degrees in diameter and is capable of covering the M31 disk with only two pointings. However, as changing filters take 30 min for HSC, it is difficult to obtain light curves in different filters simultaneously, hampering us from determining the line-of-sight extinction using the minimum light color method. Alternatively, the Large Binocular Telescope also provides a good spatial coverage (30×30 arcminutes) with its large binocular cameras. More importantly, with two 8.4 m mirrors, we can get light curves in two filters simultaneously. Please note that high cadence, wide-field imaging of M31 with large telescopes is not only useful for RR Lyra stars, but is also valuable for other short-period transients as well. It will be very useful for, e.g., microlensing studies, especially for primordial black holes and exoplanets. In this regard, Niikura et al. (2017) have carried out deep and high-cadence observations in the central region of M31. With a single exposure every 120 s and staring for almost a whole night, they obtained ~200 i-band time-series photometry with very good sky conditions (seeing <0.6 arcseconds). This data set provides a wealth of information of flare stars, eclipsing binaries, pulsating variables that are exhibiting short timescale variations. Most importantly, they only found one candidate ultra-short microlensing event (assuming that it originates from a primordial black hole), which is contradicting to the assumption that the primordial black holes could compose the majority of the dark matter. In such a scenario, we would have detected one thousand ultra-short duration microlensing events in M31

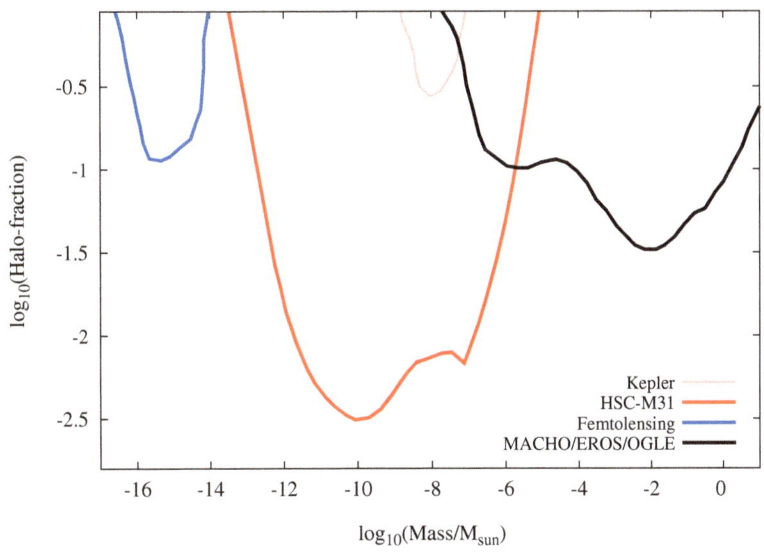

Figure 6.2. Constraints on the mass fraction of primordial black holes to dark matter in the Milky Way and the Andromeda galaxy from different primordial black hole masses. Data taken from Niikura et al. (2017). The black line marks constraints from previous microlensing surveys, i.e., MACHO, EROS, and OGLE, pointing toward the Large and Small Magellanic Clouds. The pink line marks constraints from a higher cadence survey from the *Kepler* satellite. The red line marks the most stringent constraints for primordial black holes with masses smaller than the Moon, which are derived from the HSC one-night observations of the Andromeda galaxy.

from the HSC observations of one night. In this regard, the detection of only one candidate ultra-short microlensing event translates into that primordial black holes can only make up to 0.1% of the dark matter. Thus, the primordial black holes are not the major component of dark matter, and the nature of the dark matter remain elusive (Figure 6.2).

Another future direction for M31 variable stars' studies is to study the long period variables (LPVs) in the near-infrared (see Figure 6.3). This is because there have been studies of LPVs in the Magellanic Clouds, which demonstrated a clean period–luminosity of LPVs in the K-band (Ita et al. 2004). The benefit of using LPVs instead of Cepheids as distance indicators is twofold. The first is that they are at least two magnitudes brighter than Cepheids, so they can provide a one-step distance determination out to 100 Mpc with *JWST* from space or other ELTs on the ground. The second advantage is that the LPVs are much more populated than Cepheids (~20 × more numerous), so instead of surveying a large portion of a galaxy to search for Cepheids, we can find the same number of LPVs that are required to establish the period–luminosity relation in a much smaller area. This is especially important in the era of *JWST* and ELTs, where we are limited to smaller field-of-view with the future space-based telescopes or extremely large telescopes on the ground. Currently, there has not been a dedicated survey to study the LPV population in M31, nor a dedicated infrared survey to do so. Nevertheless, with the archival time-series data, we will be able to provide a census of the LPVs in M31. Combining with the

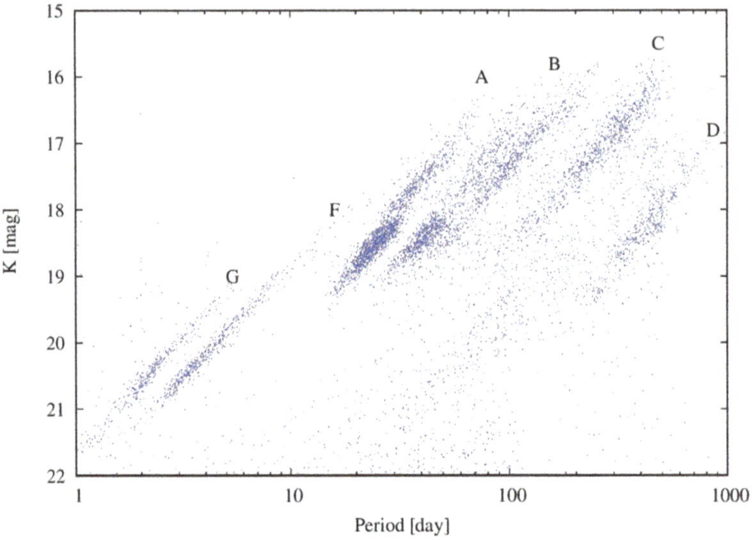

Figure 6.3. Example of infrared period–luminosity relation for Cepheids and long period variables. The data points are gathered from the infrared photometry of variables in the Large Magellanic Cloud by Ita et al. (2004), with the distance scaled to that of M31 (distance modulus of 24.4 mag). The LPVs compose four main branches, A–D, in the infrared period–luminosity diagram. The fundamental mode and first overtone Cepheids form branches F and G. The LPVs are at least two magnitudes brighter than Cepheids in the infrared and hence can be used to probe larger distances.

WIRCam archived near-infrared data, we can establish the LPV period–luminosity relation in M31, providing a firm basis for using LPVs as distance indicator in the *JWST* era.

References

Cacciari, C., & Clementini, G. 2003, in Lecture Notes in Physics 635, Stellar Candles for the Extragalactic Distance Scale, ed. D. Allion, & W. Gieren (Berlin: Springer), 105
Dolphin, A. E., Saha, A., Olszewski, E. W., et al. 2004, AJ, 127, 875
Drake, A. J., Catelan, M., Djorgovski, S. G., et al. 2013a, ApJ, 763, 32
Drake, A. J., Catelan, M., Djorgovski, S. G., et al. 2013b, ApJ, 765, 154
Ita, Y., Tanabé, T., Matsunaga, N., et al. 2004, MNRAS, 353, 705
Jeffery, E. J., Smith, E., Brown, T. M., et al. 2011, AJ, 141, 171
Kunder, A., Chaboyer, B., & Layden, A. 2010, AJ, 139, 415
Niikura, H., Takada, M., Yasuda, N., et al. 2017, arXiv:1701.02151
Pritchet, C. J., & van den Bergh, S. 1987, ApJ, 316, 517
Sarajedini, A., Mancone, C. L., Lauer, T. R., et al. 2009, AJ, 138, 184
Sesar, B., Ivezić, Ž., Grammer, S. H., et al. 2010, ApJ, 708, 717
Torrealba, G., Catelan, M., Drake, A. J., et al. 2015, MNRAS, 446, 2251

Chien-Hsiu Lee

Appendix A

ARTIS

"The only way to realize the full scientific benefit of our observations is to share the data with our competition"

—Bohdan Paczynski, 2006

While in this book I summarized some of the topics that can be done with time-domain studies of M31, there are far more sciences that are not covered here or were unexpected from the original goals of the previous M31 monitoring campaigns. As Bohdan Paczynski—the pioneer in the microlensing and all-sky variability studies—once said, we should make the data public to get the most out of it. Following this philosophy, I am compiling a master catalog and light curve database for M31 variables that are charted in the literature. Moreover, I am also collecting archival imaging data, especially from the ones by the WFCam at the 2.5 m Isaac Newton Telescope, the MegaCAM at the 3.6 m Canada–Hawaii–France Telescope, the Pan-STARRS Andromeda project, and the Palomar Transient Factory, to form a time baseline of more than one decade. The best part of the aforementioned telescopes is that they gathered data in similar filters, i.e., the g-, r-, and i-bands.

Thanks to the support of the Japan Society for the Promotion of Science, I am now reducing these archival data in a consistent manner, and most importantly, using the image subtraction method to extract pixel based light curves. I am now building a database and an online query system, called the Andromeda Real Time Inspection System (ARTIS), that will allow the users to query pixel-based light curves at any position of M31 as long as they are covered by previous time-series surveys. The pixel-based light curves are especially important if we want to investigate the variables or progenitors of the transients seen in the exquisite spatial resolution imaging from the *Hubble Space Telescope*, which are not resolvable from the ground and we can only obtain the light curves from pixel basis. The catalogs, reduced images, and the pixel-based light curves will be available at https://lchjoel1031.github.io/ARTIS/.

doi:10.1088/2514-3433/ab3639ch7

The online query system will be composed of two components, i.e., catalog-based and position-based searches. The catalog-based searches incorporate different kinds of variables (and their light curves) gathered from previous literature, including but not limited to, Cepheids, eclipsing binaries, long period variables, etc., from the DIRECT project, INT/WFCAM surveys, Pan-STARRS surveys, and so on. A selected number of catalogs and their footprint can be found in Figure A.1.

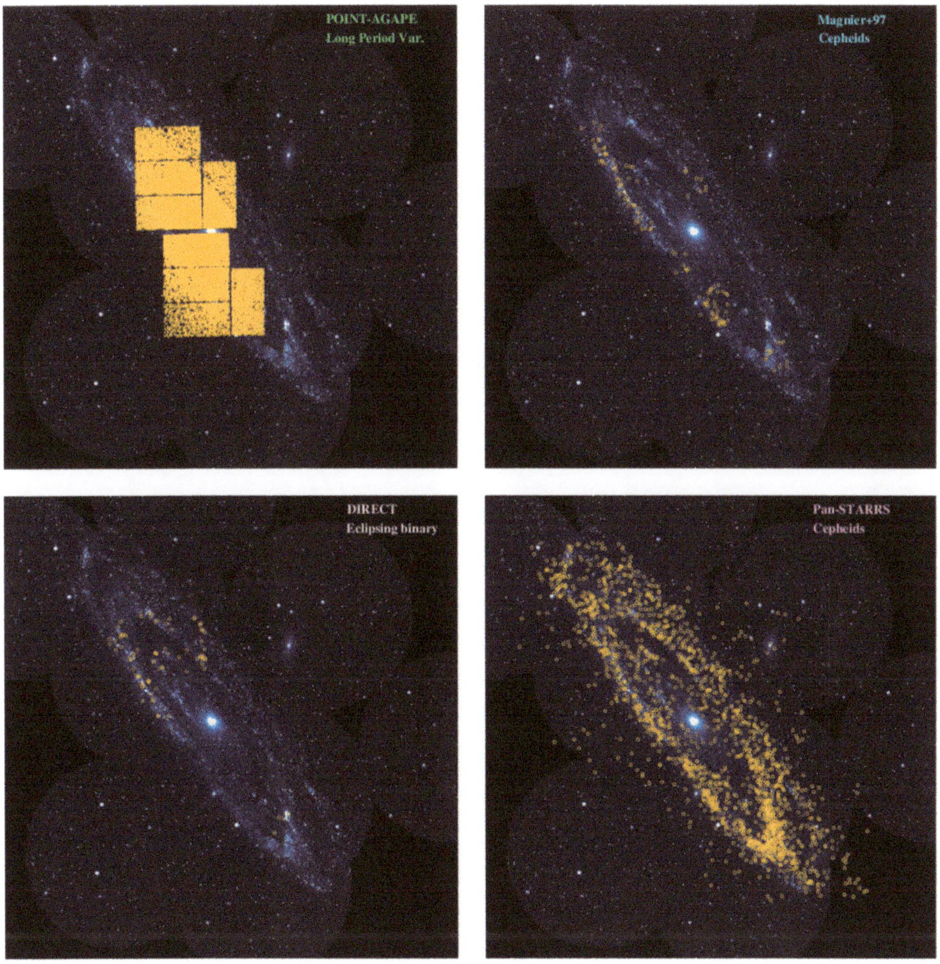

Figure A.1. Catalogs of variables (and their light curves) included in the ARTIS system.

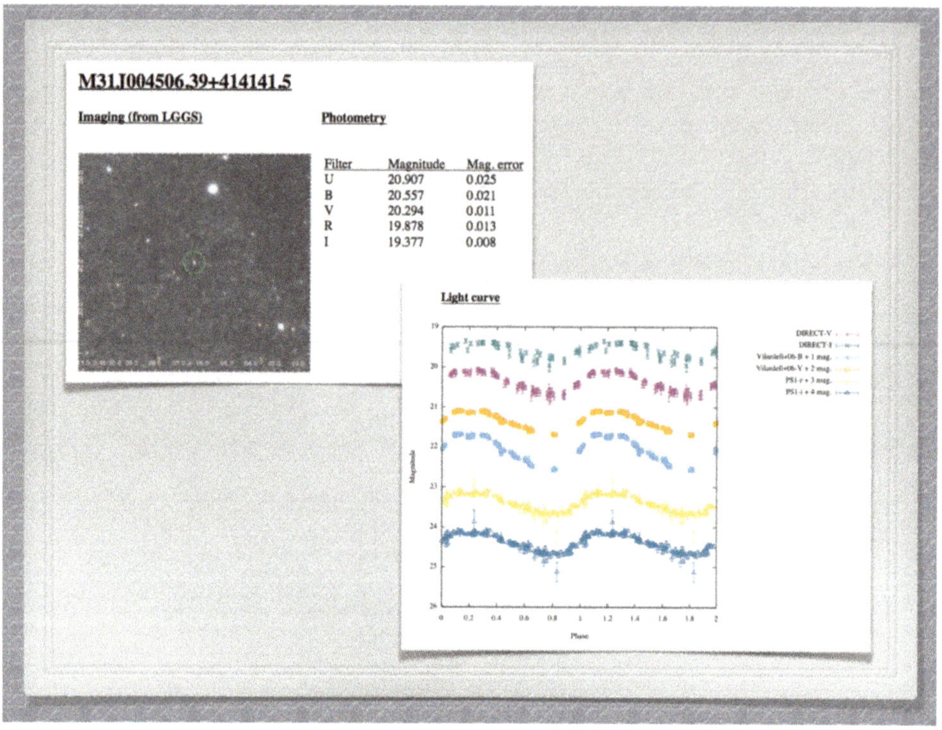

Figure A.2. An example query of light curves from ARTIS.

For the position-based searches, it will return multi-band photometry for objects within a search circle (typically <1 arcsecond) and light curves (if available from the literature). An example of the position based search can be found in Figure A.2. The pixel-based light curve search is currently under development and will be online in the near future.

Lightning Source UK Ltd.
Milton Keynes UK
UKHW021032301020
372453UK00003B/140

9 780750 313544